symposia
on
theoretical
physics

2

symposia on theoretical physics

Lectures presented at the
1964 Second Anniversary Symposium
of the Institute
of Mathematical Sciences
Madras, India

Edited by
ALLADI RAMAKRISHNAN
Director of the Institute

 PLENUM PRESS · NEW YORK · 1966

ISBN-13: 978-1-4684-7754-2   e-ISBN-13: 978-1-4684-7752-8
DOI: 10.1007/ 978-1-4684-7752-8

Library of Congress Catalog Card Number 65-21184

© 1966 Plenum Press
Softcover reprint of the hardcover 1st edition 1966
A Division of Plenum Publishing Corporation
227 West 17 Street, New York, N. Y. 10011
All rights reserved

Introduction

The second volume of this series is devoted to the Proceedings of the Second Anniversary Symposium under the chairmanship of the Niels Bohr Visiting Professor of the year – Professor L. Rosenfeld, Deputy Director of NORDITA, Copenhagen, and the Editor of *Nuclear Physics*. With particular appropriateness, the Symposium was inaugurated by the Honorable C. Subramaniam, Union Cabinet Minister, the founding father of the Institute.

The meeting was characterized by two features: (1) the enlargement of the scope of the discussions in theoretical physics, with the inclusion of many-body problems and statistical mechanics: (2) Seminars on pure mathematics, stimulated by the presence and participation of Professor Marshall H. Stone of Chicago as the First Ramanujan Visiting Professor at the Institute.

The year 1963 marked a new stage in the development of high-energy physics – the first successes of $SU(3)$ symmetry and the eightfold way had such an impact on the scientific world that the hard, unyielding domain of strong interactions was now again open to exploration. The volume opens with two significant lectures by Sudarshan and O'Raifeartaigh on fundamental problems relating to internal symmetries. The theory of Regge poles, after its initial triumph, met with rough weather, the nature and intensity of which can be realized from the series of discussions in this volume.

In statistical mechanics, we had the privilege of having with us three leading participants, Zumino of New York, Dewitt from Berkeley, and Mohling from Colorado, whose contributions to this volume summarize their academic program during their stay at our Institute.

Marshall Stone's lecture on some current trends in mathematical research was perhaps the best possible mode of initiating studies in pure mathematics at the Institute. It was followed by a systematic account of semigroup methods in mathematical physics by Bharucha-Reid. The seminar talk on the mathematical problems of cascade theory by Srinivasan is the first in a series on stochastic processes, the rest of which will follow in succeeding volumes.

As part of the surging current of scientific literature, this volume, we hope, will convey the "integrating power of mathematics"* and the "universality of physical laws."

Alladi Ramakrishnan

---

* I am indebted for this phrase to Professor M. J. Lighthill, who referred to the role of mathematics in his inaugural lecture entitled "Waves in Fluids" on assuming the Royal Society Research Professorship in 1965.

Contents

Contents of Other Volumes

Origin of Internal Symmetries

E. C. G. SUDARSHAN

UNIVERSITY OF ROCHESTER
Rochester, New York

1. INTRODUCTION

Symmetry groups in physics seem to belong to two classes: the so-called relativity (or frame) groups, which may be called the *external symmetry groups*, defined by the geometric relations between "inertial" systems for which the laws of physics are the same, and the *internal symmetry groups*. We call the symmetry "internal" because we see only its manifestations; there is no primitive geometric characterization of the symmetry group from any fundamental dynamic principle. We shall try to see to what extent a dynamic principle can be expected to generate a symmetry group.

In this connection, two sets of quantum numbers can be distinguished—the additive quantum numbers (such as the third components of $\vec{J}$ and $\vec{T}$), which are algebraically additive, and the nonadditive ("vector") quantum numbers (such as the total angular momentum $\vec{J}$, total isotopic spin $\vec{T}$, etc.), which obey vector laws of addition and multiplication. One fact worth recalling is that the irreducible representations of a compact group are finite dimensional and are equivalent to unitary representations.

We naturally ask about the properties of particles in interaction. Suppose, for example, we consider the following (virtual) reaction:

$$N \rightarrow N + \pi$$

From the $(NN\pi)$ vertex, we can write the invariant interactions (by using the Clebsch–Gordan coefficients) and obtain the following relationships between the various $(NN\pi)$ coupling constants (g) and among the various virtual transition probabilities:

1

$$g_{pp\pi^0} = -g_{nn\pi^0}$$

$$= \frac{1}{\sqrt{2}} g_{pn\pi^+}$$

$$= \frac{1}{\sqrt{2}} g_{np\pi^-}$$

$$\Gamma(n \rightarrow n + \pi^0) = \Gamma(p \rightarrow p + \pi^0) \qquad (1)$$

$$= \frac{1}{2} \Gamma(p \rightarrow n + \pi^+)$$

$$= \frac{1}{2} \Gamma(n \rightarrow p + \pi^-)$$

where p and n refer to the proton and neutron, respectively. From these we conclude that for the total widths

$$\Gamma(n \rightarrow \text{any particle}) = \Gamma(p \rightarrow \text{any particle})$$

We know that the multiplet structures displayed by the known particles are consequences of the (postulated) existence of an internal symmetry. We therefore ask whether the existence of the multiplet structure conversely implies an internal symmetry.

Recently, there have been a good many attempts to explain the internal symmetry by some direct dynamic calculations. If we start with a multiplet of N vector mesons of equal masses and assume that the interactions among these vector mesons are essentially trilinear in character, we can make a dynamic scheme in terms of a straight-forward and self-consistent bootstrap mechanism between these (equally massive) vector mesons. One such attempt was made by Capps,[1] who found that the interactions among these N equally massive vector mesons obey unitary symmetry (i.e., invariance under the group SU_3). Capps was surprised to find this relationship between unitary symmetry and a self-consistent bootstrap calculation. It looked as though unitary symmetry could be derived from first principles. However, it is possible that the symmetry would have emerged from the assumption of the existence of a multiplet degenerate in mass before the interaction and the postulate that this multiplet structure is preserved even in the presence of interactions, so that the particles exhibit the same mass degeneracy even in the presence of the interaction if they have equal masses. Such arguments have been used by Sakurai,[2] who tried to prove that the emergence of the symmetry is not a consequence of a sophisticated dynamic calculation, but rather the immediate consequence of the assumptions:

1. Equality of masses of the particles.
2. Existence of a degenerate multiplet before the interaction.
3. The presence of interaction not altering the equality of masses (or the multiplicity of the particles).

He directly shows, as an example, that if we equate the contributions of certain self-energy diagrams we arrive at the required symmetry.

Thus, if we equate the second-order self-energy of the nucleons (assuming equality of masses before the interaction) as shown in Fig. 1a and that of the pions as shown in Fig. 1b, we obtain

$$2g^2_{pp\pi^0} = 2g^2_{nn\pi^0} = g^2_{pn\pi^+} = g^2_{np\pi^-}$$

In this calculation the multiplets are treated "on the same footing," and the "total width" for each component comes out to be equal. By taking fourth-order diagrams also, we can further deduce

$$g_{pp\pi^0} = -g_{nn\pi^0}$$

Thus, the symmetries may well be explained if we assume the equality of masses and multiplicity of the particles and postulate that these properties remain unchanged even in the presence of interaction.

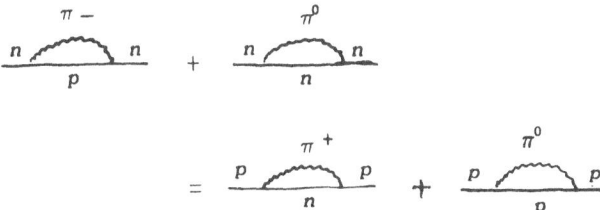

Fig. 1

2. GAUGE FIELDS

The consequence of the existence of symmetries and the postulated invariance of interactions under the gauge transformations of the second kind is the existence of vector gauge fields coupled linearly to conserved quantities (such as electric charge, etc.).

Unlike the electromagnetic field, which by itself is neutral and interacts only with charged fields (and is thus coupled to the electric current), gauge vector fields may themselves carry the properties. The isospin gauge field, for instance, itself carries the properties of isospin, and it is hence nonlinearly self-coupled. We may even consider a situation in which the gauge vector field alone carries isotopic spin and is consequently self-coupled. Thus, if we can write $L = j^\mu A_\mu$ for the Lagrangian of the electromagnetic interaction, where A_μ is the electromagnetic field and j^μ is the current to which it is coupled, what can we write for the Lagrangian of the interaction of the gauge vector field? Since the gauge vector field is coupled to itself, we naturally expect that the interaction can be written as a product of these fields B. Then how many B can enter the product? The simplest possibility (which we may take to be basic) is the trilinear interaction between vector particles. This is because the current is bilinear in B field and coupled to another B field, making a trilinear vertex.

Cutkosky[3] has given a simple model in which he assumes that there are a number (N) of vector mesons which have the same mass, i.e., he assumes a multiplet structure. Then, with a number of additional plausible assumptions, he shows that a Lie group could be associated with these particles. The assumptions made are:

1. The vector mesons arise as self-consistent bound states of pairs of vector mesons.

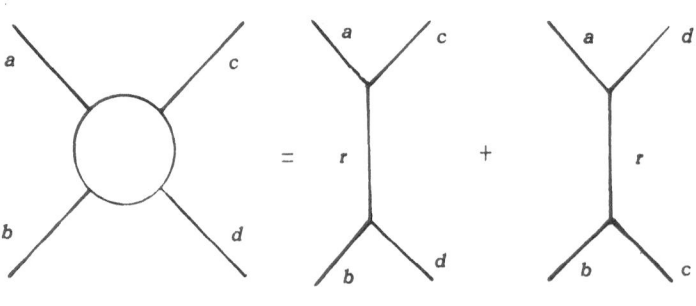

Fig. 2

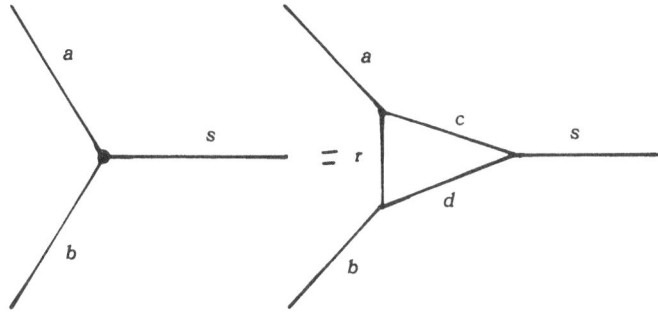

Fig. 3

2. The binding force is mediated by the exchange of single vector mesons; for example, the long-range part of the force is given by the one-particle exchange diagrams shown in Fig. 2.

3. The renormalized coupling constants are well approximated by the simplest irreducible vertex part, with the "bare-coupling constants" set equal to zero, as shown in Fig. 3.

4. Parity is conserved in strong interactions, and strong interactions are invariant under charge conjugation.

5. Electric charge is conserved.

6. The dependence of the vertex on the internal labels F_{abc} is antisymmetric in all pairs of indices.

If we represent the particles by real vector fields $B_\mu(\mu = 1 \ldots N)$, the invariant interaction has the form

$$F_{abc} B_a B_b B_c$$

and F_{abc} is antisymmetric. We then look for the eigenfunctions of F_{abc}. The Born-approximation scattering amplitude is proportional to

$$V_{ab,cd} = (F_{adr} F_{bcr} - F_{acr} F_{bdr}) \tag{2}$$

corresponding to the two diagrams in Fig. 2 and taking into account the antisymmetric nature of F.

Since all the particles which are together, and also all the exchanged particles, have the same mass (which we have normalized to unity), it is clear that we can obtain N degenerate bound states only if V has N degenerate eigenvalues. Also, the F themselves must be eigenvectors of V, in view of postulate (3):

$$V_{ab,cd} F_{cdS} = \lambda F_{abS} \tag{3}$$

where $\lambda > 0$. Since F_{abr} can be interpreted as the internal "wave function" of the particle considered as a bound state of the two particles (a, b), we may normalize F as

$$F_{abr} F_{bas} = \delta_{rS} \tag{4}$$

If the model is self-consistent, we must see that no particle comes out with a mass less than the mass with which we started. Therefore, of all the antisymmetric eigenvectors of V ($V\psi_i = \lambda_i \psi_i$), we should allow only those for which

$$\lambda > \lambda_i \tag{5}$$

that is, we must require that no other vector particles which have a lower mass than the N we started with should arise from the potential; otherwise, the model will not be self-consistent. We then proceed to determine F, satisfying equations (3), (4), and (5). We have from the definition of trace

$$\mathrm{Tr}\, V^2 = N\lambda^2 + \sum_i \lambda_i^2 \tag{6}$$

The explicit form of V given by equation (2), together with equations (3) and (4), can be used to calculate the alternative expression

$$\mathrm{Tr}\, V^2 = 2N - N\lambda \tag{7}$$

Therefore, it follows that

$$\lambda(\lambda + 1) = 2 - \frac{1}{N} \sum \lambda_i^2 \tag{8}$$

so that

$$\lambda \leq 1 \tag{9}$$

The equality holds when $\sum \lambda_i^2 = 0$. Under the orthogonal transformation

$$B'_a = 0_{ab} B_b$$

equations (3) and (4) are covariant. For infinitesimal transformations of 0_{ab}:

$$0_{ab} = \delta_{ab} + i\epsilon^\alpha G_{ab}^\alpha$$

the F_{abc} transform according to

$$F'_{abc} = F_{abc} + i\epsilon^\alpha f_{abc}^\alpha$$

$$f_{abc}^\alpha = F_{rbc} G_{ra}^\alpha + F_{arc} G_{rb}^\alpha + F_{abr} G_{rc}^\alpha \tag{10}$$

Cutkosky makes assumption (5), that the interactions satisfy a nontrivial additive conservation law (say, conservation of charge at each vertex). This requirement simplifies the analysis, because F_{abc} are invariant under gauge transformations of the first kind if we hold this assumption. (If we had assumed the existence of l independent additive conservation laws, F_{abc} would have been invariant under an l parameter Abelian subgroup of 0_N.) If we denote by G_{ab}^A a generator of the Abelian gauge transformation, then

$$F_{rbc} G_{ra}^A + F_{arc} G_{rb}^A + F_{abr} G_{rc}^A = 0 \tag{11}$$

If we multiply equation (11) by F_{bad}, and use the fact that F is antisymmetric, we get

$$G_{cd}^A = V_{cd,ab} G_{ab}^A \tag{12}$$

The generators G_{ab}^A are eigenvectors of V with unit eigenvalue. Hence, from equations (5) and (8) it follows that all $\lambda_i = 0$; consequently, the completeness of the eigenvectors of V then allows us to write

$$V_{ab,cd} = F_{abr} F_{dcr}$$

which can be written as

$$F_{abr} F_{cdr} + F_{bcr} F_{adr} + F_{car} F_{bdr} = 0 \tag{13}$$

using the definition of V given by equation (2). Equations (13) and (15) are the necessary and sufficient conditions for the F_{abc} to be the structure constants of a compact Lie group. The association is necessarily with a particular representation of the group, the adjoint representation.

Quite recently, Weinberg[4] stated that charge conservation seems to play a crucial role in generating continuous symmetries. He observed that any discrete (or continuous) symmetry arising from dynamics will always be transmuted into a full-fledged Lie group by the condition of charge conservation, provided that the electric charge operator is not invariant under the original symmetry. If U is any member of the group of physical symmetries, then also is

$$U^{-1}[\exp{(i\theta Q)}] U = \exp{[i\theta(U^{-1}QU)]}$$

where Q is the charge operator. Hence, both Q and $U^{-1}QU$ belong to the Lie algebra of the physical symmetry group. When U does not commute with Q, then we can generate a larger symmetry group.

However, it seems to be really possible that we may relax the condition of charge conservation in obtaining internal symmetries. In a model calculation,[5] we find that charge conservation comes out naturally and need no longer be imposed. This is a direct self-energy calculation such as that of Sakurai[2] with no condition on electric charge conservation. We can also try to relax the condition of charge conservation from Cutkosky's calculation.[3]

Let us first consider the modified Cutkosky model. The potential established by the exchange of single vector mesons between pairs of vector mesons (a, b) and (c, d) in the Born approximation can be written as

$$V_{ab,cd} = (F_{acr}F_{bdr} - F_{adr}F_{bcr}) \qquad (14)$$

corresponding to the two diagrams given in Fig. 2. This potential to a very good degree of approximation is also the scattering amplitude which will contain pole terms corresponding to single particle exchanges to the direct channel. We omit higher intermediate particle contributions to the scattering amplitude, i.e., in the pole approximation

$$\text{Scattering amplitude} = \sum_r \frac{F_{abr}F_{cdr}}{s - \mu^2} \qquad (15)$$

which corresponds to the pole diagram shown in Fig. 4. Here s is the total center of mass energy squared and μ is the mass of the particle exchange. Thus, the potential

$$\begin{aligned} V_{ab,cd} &= -F_{acr}F_{bdr} + F_{adr}F_{bcr} \\ &= \chi F_{abr}F_{cdr} \end{aligned} \qquad (16)$$

where χ is a constant which does not depend on r. (This equation can of course be never true as it stands; the left-hand side terms have poles in the momentum transfer variables t and u, while the right-hand side term has a pole in the energy variable. However, if we iterate either the left-hand (or the right or both!) terms, we will change their dependence on s, t, u. But it is possible that their dependence on the internal labels a, b, c, d is unaltered. We consider this possibility here only, and after the solution is obtained we can in fact verify that a "horizontal" iteration leaves the a, b, c, d dependence essentially unchanged, since a, b, c, d turns out by virtue of equation (18) to be a constant multiple of a projection matrix. We short-circuit these essential dynamic points in the sequence of arguments in the text. Equation (16) is an identity which we can write as

$$\chi F_{abr}F_{cdr} - F_{adr}F_{bcr} + F_{acr}F_{bdr} \equiv 0 \qquad (16a)$$

Making all permutations b, c, d in equation (16a), taking into account the antisymmetric nature of the F, and adding all such equations, we can obtain the equation

$$(2 - \chi)[F_{abr}F_{cdr} + F_{acr}F_{bdr} + F_{adr}F_{bcr}] \equiv 0 \qquad (17)$$

But χ is a function of the invariant energy and momentum transfer

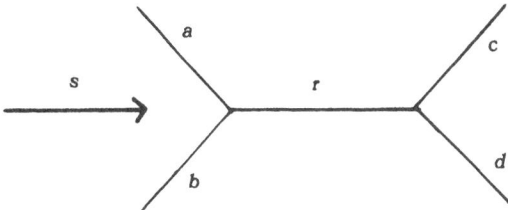

Fig. 4

variables, and therefore almost everywhere the factor $(2 - \chi)$ is nonzero. Hence

$$F_{abr}F_{cdr} + F_{acr}F_{bdr} + F_{adr}F_{bcr} \equiv 0 \qquad (18)$$

Equation (18) is just the Jacobi identity obeyed by the structure constants of a Lie group. Consequently, from the group property, there should be at least one Abelian subgroup corresponding to conservation of an additive quantum number, which we choose to be the charge. Thus, charge conservation comes out of the calculation.

Next we shall see that charge conservation can in fact be relaxed from Sakurai's calculation of self-energy contributions.[5] Consider now the self-energy diagrams of pion and nucleons shown in Fig. 4. Here the labels $\alpha, \beta \ldots$ correspond to the mesons and r, s, t correspond to nucleons. Let us at this stage state the generalized Smushkevich principle: "The (dressed) propagators of the component fields of a multiplet are the same."

A more useful and (possibly) equivalent statement of the Smushkevich principle is the following: "Topologically identical self-energy diagrams should give equal contributions to the propagators of component fields of a multiplet."

Suppose we write down the nucleon self-energy contribution from Fig. 5. This is equal to

$$\sum_{\alpha, t} g_{rt\alpha} g_{ts\alpha} \qquad (19)$$

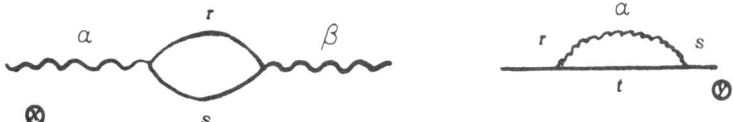

Fig. 5

where $g_{rt\alpha}$ is the coupling constant corresponding to the vertex $(rt\alpha)$. Let us define the matrices C by

$$g_{rt\alpha} = C_{rt}^{\alpha} \qquad (20)$$

so that equation (19) becomes

$$\begin{aligned}
\sum_{\alpha,t} g_{rt\alpha} g_{ts\alpha} &= \sum_{\alpha,t} C_{rt}^{\alpha} C_{ts}^{\alpha} \\
&= \sum_{\alpha} (C^{\alpha} C^{\alpha})\ rs\text{-th element} \qquad (21) \\
&= A\delta_{rs}
\end{aligned}$$

by Smushkevich's principle, where A is a constant independent of r and s. Thus

$$\sum_{\alpha} (C^{\alpha} C^{\alpha}) = AI \qquad (22)$$

Similarly, the contribution from Fig. 4 to the pion self-energy is equal to

$$\begin{aligned}
\sum_{r,s} g_{rs\alpha} g_{sr\beta} &= \sum_{r,s} C_{rs}^{\alpha} C_{sr}^{\beta} \\
&= \sum_{r} C_{r}^{\alpha} C_{r}^{\beta} \\
&= \text{Tr}(C^{\alpha} C^{\beta}) \qquad (23) \\
&= B\delta_{\alpha\beta}
\end{aligned}$$

by the Smushkevich principle, where B is a constant independent of α and β. Thus

$$\text{Tr}\, C^{\alpha} C^{\beta} = B\delta_{\alpha\beta} \qquad (24)$$

We can easily show that $3B = 2A$, but we shall not use this result. Therefore, our task is to find the matrices C which satisfy equations (22) and (24). Since the pion fields retain their properties under real orthogonal transformations:

$$C^{\alpha} \longrightarrow \sum_{\beta} 0^{\alpha\beta} C_{\beta} \qquad (25)$$

and the nucleon fields retain their properties under arbitrary unitary transformation

$$C^{\alpha} \longrightarrow UC^{\alpha} U^{-1}$$

we can show that the C are the matrices that we want, namely, τ_1, τ_2, and τ_3. It is always possible, by an orthogonal transformation, to have two of the matrices C_1, C_2, and C_3 be traceless. It may be noted that the C are Hermitian, i.e., we can have

$$\begin{aligned}
\text{Tr}\, C_1 &= 0 \\
\text{Tr}\, C_2 &= 0
\end{aligned} \qquad (26)$$

Since any traceless Hermitian matrix can be made a multiple of τ_1 by a unitary transformation, let us choose C_1. [Incidentally, it is clear that $\mathrm{Tr}\,(\tau_1\,\tau_1) = 2$ and therefore satisfies equation (24) if we choose $B = 2$.] That is,

$$C_1 = \tau_1 \tag{27}$$

[Otherwise, we get $C_1 = (B/2)^{1/2}\,\tau_1$.] C_2, which is to be traceless and Hermitian, can be expanded as

$$C_2 = a\tau_1 + b\tau_2 + c\tau_3 \tag{28}$$

where a, b, and c are real coefficients. In this, we exclude the unit matrix by virtue of condition (26). Let us now use equations (22) and (24) to evaluate the coefficients a, b, and c.

$$\mathrm{Tr}\,C_2\,C_2 = 2 = 2(b^2 + c^2)$$

therefore

$$b^2 + c^2 = 1$$

$$\mathrm{Tr}(C_1\,C_2) = 0 = 2a \Longrightarrow a = 0 \tag{29}$$

We can choose $b = \cos\theta$ and $a = -\sin\theta$. Thus

$$C_2 = \tau_2 \cos\theta - \tau_3 \sin\theta \tag{30}$$

By a rotation around the τ_1 axis with our choice of C_1 unaltered, we can get C_2 as

$$\exp\left(\tfrac{1}{2}\boldsymbol{\tau}_1\cdot\boldsymbol{\theta}\right)C_2 \exp\left(-\tfrac{1}{2}\boldsymbol{\tau}_1\cdot\boldsymbol{\theta}\right) = \tau_2 \tag{31}$$

Thus, having found C_1 and C_2, let us now find C_3. In general

$$C_3 = d\tau_1 + e\tau_2 + f\tau_3 + gI \tag{32}$$

The unit matrix is included since we do not need C_3 to be traceless. Coefficients d, e, f, and g are real. We now use equations (22) and (24) and the choice of C_1 and C_2 to find d, e, f, and g:

$$\mathrm{Tr}\,C_1\,C_3 = 0 \Longrightarrow d = 0$$

$$\mathrm{Tr}\,C_2\,C_3 = 0 \Longrightarrow e = 0$$

$$\mathrm{Tr}\,C_3\,C_3 = 2 = 2(f^2 + g^2)$$

$$f^2 + g^2 = 1$$

therefore

$$C_3 = f\tau_3 + gI \tag{33}$$

If we use equation (22)

$$C_1\,C_1 + C_2\,C_2 + C_3\,C_3 = AI$$

so that

$$1 + 1 + f^2 + g^2 + 2fg\tau_3 = AI$$

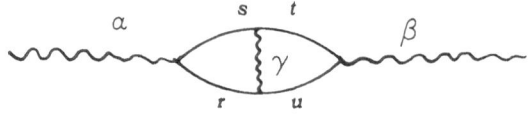

Fig. 6

Since each term on the left-hand side has to be a multiple of the unit matrix, the only term which is not a multiple of identity should vanish:

$$fg = 0 \Longrightarrow f = 0 \quad \text{or} \quad g = 0$$

Thus we are left with two choices:

$$C_3 = \tau_3 \quad \text{or} \quad C_3 = I \tag{34}$$

In the following, we will show that the choice (τ_1, τ_2, I) for the C contradicts the Smushkevich principle given in equation (24). If we go to the next higher order self-energy diagram for the meson (Fig. 6), the contribution is

$$\sum_{\substack{r,s,t,u \\ \gamma}} g_{rs\alpha} g_{str} g_{ut\beta} g_{ur\gamma}$$

$$= \sum_{\substack{r,s,t,u \\ \gamma}} C_{rs}^{\alpha} C_{st}^{\gamma} C_{tu}^{\beta} C_{ur}^{\gamma}$$

$$= \sum_{\gamma} \mathrm{Tr}(C^{\alpha} C^{\gamma} C^{\beta} C^{\gamma}) \tag{35}$$

$$= B' \delta_{\alpha\beta}$$

by the Smushkevich principle, where B' is another constant independent of α and β.

If we choose the basis (τ_1, τ_2, I) for the C, then for $\alpha = \beta = 1$ or 2 we really get

$$\sum_{\gamma} \mathrm{Tr}(C^{\alpha} C^{\gamma} C^{\beta} C^{\gamma}) = 2$$

and hence equation (24) is satisfied. However, for $\alpha = \beta = 3$, we get

$$\sum_{\gamma} \mathrm{Tr}(C^{\alpha} C^{\gamma} C^{\beta} C^{\gamma}) = \mathrm{Tr}(I\tau_1 I\tau_1)$$
$$+ \mathrm{Tr}(I\tau_2 I\tau_2)$$
$$+ \mathrm{Tr}(IIII)$$
$$= 2 + 2 + 2$$

and thus equation (35) is not satisfied. Hence, we are left with the unique choice (τ_1, τ_2, τ_3) for the C. The ambiguity about the sign (i.e., choice $\pm\tau_1, \pm\tau_2, \pm\tau_3$) is there, but this can be eliminated by a redefinition of the pion fields. Thus we end up with a $(\boldsymbol{\tau} \cdot \boldsymbol{\phi})$ interaction for the π-N system. In other words, the interaction is such that the charge is conserved.

Therefore, we have seen that the idea of charge conservation can be suppressed in both Cutkosky's and Sakurai's calculations. The question is whether we can generalize this statement to all interactions. The answer seems to be no, since we have a counterexample for the ($\Sigma\Sigma\pi$) system. A similar fate befalls Sakurai's method, which includes electric charge conservation for higher isospins.

It appears then that no particular axiom is crucial in the "derivation of an internal symmetry," since Sakurai's calculation relaxes self-consistency, and we have shown that even charge conservation can be relaxed. The question is how many axioms we need in order to specify the internal symmetry. The point of view advanced here is that the Smushkevich principle, when augmented by suitable auxiliary restrictions, is the really basic one in understanding the internal symmetry. The origin of internal symmetries is still open, although we have some indication as to the direction in which a correct solution may lie. These demonstrations have now been extended[5] to a variety of systems—in particular, to a system of two multiplets of n components each coupled to a multiplet of ($n^2 - 1$) members to deduce invariance under the special unitary group $SU(n)$, again without assuming electric charge conservation.

REFERENCES

1. R.H. Capps, *Phys. Rev. Letters* **10**: 312 (1963).
2. J.J. Sakurai, *Phys. Rev. Letters* **10**: 446 (1963).
3. R.E. Cutkosky, *Phys. Rev.* **131**: 1888 (1963); E.C.G. Sudarshan, *Phys. Letters* **9**: 286 (1964).
4. S. Weinberg, "On the Derivation of Internal Symmetries," University of California preprint.
5. E.C.G. Sudarshan, L. O' Raifeartaigh, and T.S. Santhanam, *Phys. Rev.* **136B**: 1092 (1964).

Construction of the Invariants of the Simple Lie Groups

L. O'RAIFEARTAIGH

DUBLIN INSTITUTE FOR ADVANCED STUDIES
Dublin, Ireland

To introduce the problem which I should like to consider in this talk, let me first make some remarks about the three-dimensional rotation group O_3, with which we are familiar. For this group, the infinitesimal generators T_i, $i = 1, 2, 3$, satisfy the commutation relations

$$[T_i T_j] = iT_k \qquad (i, j, k \text{ cyclic}) \tag{1}$$

and there exists a polynomial in the T_i, namely,

$$T^2 = T_1^2 + T_2^2 + T_3^2 \tag{2}$$

which has the property that it commutes with *all* of the infinitesimal generators:

$$[T^2, T_i] = 0 \qquad (i = 1, 2, 3) \tag{3}$$

Furthermore, it is the only independent polynomial in the T_i with this property, since any other polynomial P satisfying equation (3) can be expressed as a polynomial in T^2:

$$P = P(T^2) \tag{4}$$

Our problem is simply the generalization of this, as follows:
Let

$$[X_\alpha, X_\beta] = C_{\alpha\beta}^\gamma X_\gamma \qquad (\alpha, \beta, \gamma = 1, \ldots, r) \tag{5}$$

be any semi-simple Lie group, of order r and rank l, with structure constants $C_{\alpha\beta}^\gamma$. (For our purposes there need be no distinction made between simple and semi-simple Lie groups, since the latter can always be expressed as a direct product of the former.) The problem is to find

15

all the polynomials $F(X_\alpha)$, or rather all the *independent* polynomials $F(X_\alpha)$, such that

$$[F(X_\alpha), X_\beta] = 0 \qquad (\beta = 1, \ldots, r) \qquad (6)$$

that is, to find all the independent *invariants* $F(X_\alpha)$.

Now, why is this problem interesting? First, from the mathematical point of view, it is obvious that the quantities with the property (6) are going to play a very fundamental role. Let me give one example, however. For the group O_3, the representations can be labeled by j, the highest eigenvalue of T_3. As is well known, $j = \frac{1}{2}, 1, \frac{3}{2}$ gives the possible representations. But the representations can also be labeled according to the values of T^2, which is constant for each representation and in fact equal to $j(j + 1)$. In the same way, for the general group, the representations are labeled by the highest weights, $j_1, j_2, \ldots, j_l$, but if we had the independent $F(X_\alpha)$ of equation (6), we would have an alternative method of labeling the representations.

To see why this problem is interesting from the physical point of view, let us also make an analogy to O_3. For isotopic spin, we take the various "elementary" particles, assign them to various representations of O_3, and, having done so, express the charge independence of the strong interactions by saying, essentially, that the strong S-matrix S is a function of T^2 only, and not of the individual T_i:

$$S = S(T^2) \qquad (7)$$

Similarly, we express the idea of "higher symmetry" by taking a group of higher rank, SU_3 for example, assigning the various "particles" and resonances to the various representations of this, and then postulating that the strong, or very strong, interaction S-matrix is a function of the X_α only via the invariants $F(X_\alpha)$ of equation (6).

So, that is our problem, and this is a brief indication of why it is of interest (at least to me).

I should mention at this stage that we would like to be able to solve this problem by infinitesimal means, if possible. This is because, for the physicist, what is important is not so much the Lie group itself, but the complex Lie algebra.

The question now is: How are we going to approach the problem? What we shall do is to adopt the infinitesimal approach initiated by Casimir, who discovered that T^2 of O_3 is a special case of the so-called Casimir operator

$$C_\alpha = g_{\alpha\beta} X^\alpha X^\beta = C_{\sigma_1\alpha}^{\sigma_2} C_{\sigma_2\beta}^{\sigma_1} X^\alpha X^\beta \qquad (8)$$

which (as can easily be verified from the Jacobi identity) is an invariant for any Lie group. This operator was later generalized by Racah to

$$C_n = C^{\sigma_n}_{\sigma_1\alpha} C^{\sigma_1}_{\sigma_2\beta} \ldots C^{\sigma_{n-1}}_{\sigma_n\gamma} X^\alpha X^\beta \ldots X^\gamma \qquad (9)$$
$$\underleftarrow{\quad n \text{ terms} \quad}$$

which (again easily verified) are invariants. However, we find that in the C_n we have *not* included *all* the invariants of the group. But the possibility of generalizing (9) presents itself. This is seen by noting that if we denote the *adjoint* representation of the group by $\overset{A}{X}_\alpha$, we have

$$\overset{A}{X}_\alpha = C^\lambda_{\mu\alpha} \qquad (10)$$

where the λ, μ are to be regarded as matrix indices, and so

$$C_n = (\text{Sp } \overset{A}{X}_\alpha \overset{A}{X}_\beta \ldots \overset{A}{X}_\gamma) X^\alpha X^\beta \ldots X^\gamma \qquad (11)$$

This immediately suggests using, instead of the adjoint representation $\overset{A}{X}_\alpha$, any given representation $\hat{X}_\alpha$ and forming the quantities

$$I_n = (\text{Sp } \hat{X}_\alpha \hat{X}_\beta \ldots \hat{X}_\gamma) X^\alpha X^\beta \ldots X^\gamma \qquad (12)$$

We hope that these will be invariants and that they will be *all* the invariants. And so they are. The only trouble is that we have here *un embarras de richesses*. We have a double infinity of invariants, since, first, we can take any representation $\hat{X}_\alpha$ and, second, we can take the representations of all orders for any given $\hat{X}_\alpha$. Our problem will be to pick out the independent ones. Thus we have three problems:

1. To prove that the I_n are invariant.
2. To prove that they include all the invariants, or at least all the independent invariants.
3. To pick out the independent ones.

To these three problems I might add a fourth:

4. To obtain the independent invariants in a form which will enable us to write them simply and explicitly.

First, let us tackle (1), because this problem we can handle in a rather simple way by defining for any X_α the quantity

$$U = 1 + \epsilon_\alpha X_\alpha \qquad (13)$$

where the ϵ_α are r arbitrary (but small) numbers. Then we observe how the X_β vary under a U-transformation: that is, we have

$$\overset{\displaystyle C^\gamma_{\alpha\beta} X_\gamma}{\underset{|||}{}}$$
$$U X_\beta U^{-1} = X_\beta + \epsilon_\alpha [X_\alpha, X_\beta] = (\delta^\gamma_\beta + \epsilon_\alpha C^\gamma_{\alpha\beta}) X_\gamma = a^\gamma_\beta X_\gamma \qquad (14)$$

which means that, under U, the X_β transform as a vector, with coefficients a_{β}^{γ}. Of course, the a_{β}^{γ} are not arbitrary, but are defined in terms of the ϵ_α and the $C_{\alpha\beta}^{\gamma}$.

We can now proceed to the proof of invariance. In fact, we have

$$
\begin{aligned}
UI_n U^{-1} &= (\text{Sp } \hat{X}_\alpha \hat{X}_\beta \ldots \hat{X}_\gamma) UX^\alpha X^\beta \ldots X^\gamma U^{-1} \\
&= (\text{Sp } \hat{X}_\alpha \hat{X}_\beta \ldots \hat{X}_\gamma) a_\lambda^\alpha a_\mu^\beta \ldots a_\nu^\gamma X^\lambda X^\mu \ldots X^\nu \\
&= (\text{Sp } \hat{X}_\alpha \hat{X}_\beta \ldots \hat{X}_\gamma a_\gamma^\alpha a_\mu^\beta \ldots a_\nu^\gamma) X^\lambda X^\mu \ldots X^\nu \\
&= (\text{Sp } \hat{U} \hat{X}_\lambda \hat{X}_\mu \ldots \hat{X}_\nu \hat{U}^{-1}) X^\lambda X^\mu \ldots X^\nu \\
&= (\text{Sp } \hat{X}_\lambda \hat{X}_\mu \ldots \hat{X}_\nu) X^\lambda X^\mu \ldots X^\nu \\
&= I_n
\end{aligned}
\tag{15}
$$

Hence, I_n commutes with U and, since the ϵ_α are arbitrary,

$$
[I_n, X_\alpha] = 0 \tag{16}
$$

as required. So that settles problem (1).

With regard to problem (2), I will not be able to prove here that we have in the I_n all the independent invariants. I will simply say that there are two proofs, one of which is global (i.e., the properties of the group as a whole are used) and the other infinitesimal. Later, when I have listed the independent invariants, I shall try to make it plausible that they do form a complete set. I should mention, perhaps, that both the proofs just mentioned are used only *after* the independent invariants among the I_n are first written down.

This brings us to problem (3). Its solution was first written by Racah,[1] and I will here give the solution only for the four general classes of Lie groups A_l, B_l, C_l, and D_l. There are no particular special features in the cases of the five special groups G_2 F_4, etc.

The independent invariants are listed in Table I, and we can see at once the special role played here by the self-representation. In fact, were it not for the appearance of the one "spinor invariant" in the case of D_l, it would be tempting to say that our method is unnecessarily complicated and that the result could have been predicted directly from the definition of the linear groups. If we ignore the "spinor invariant," we see that for these invariants the self-representation has taken over completely from the adjoint representation used to form the generalized Casimir invariants C_n of equation (9).

Our next problem is to try to make plausible our results that the invariants listed in Table I are the only independent invariants and that they form a complete set. That they are the only independent ones may be seen roughly as follows: Out of the self-representations of

A_l, B_l, C_l, and D_l it is possible to build all of the tensor representations by taking direct products of the self-representation with itself. Since the tensor representations are "built up" in this way from the self-representations, it seems quite plausible that the invariants formed with them will be "built up" out of the invariants formed with the self-representation. And it turns out that this is the case.

But what about the spinor representations? For these, we can invert the procedure (to a certain extent!). Since the tensor representations can be obtained from direct products of the spinor representations, we find that the "tensor invariants" are functions of the "spinor invariants." But it turns out that (with one exception) these functional equations can be solved, yielding the "spinor invariants" as functions of the "tensor invariants" and so of the "self-representation invariants." The exception is just the peculiar l-th order spinor invariant in the case of D_l.

Details of the above arguments will be published later,[3] together with the infinitesimal proof of the completeness of the set of invariants listed above. (The global proof has already been given by Racah.[1,2])

To make it plausible that our set of invariants is *complete*, we note that in each representation the invariants are simply multiples of the unit matrix and hence are functions of the representations only. But each representation is specified by its highest weight—a set of l numbers

Table I. Independent Invariants

Group	Description (linear realization)	Order of invariants	Representation used to form it
A_l	All unimodular $(l+1)(l+1)$ matrices	$2, 3, 4 \ldots l+1$	Self-representation
B_l	All orthogonal $(2l+1)(2l+1)$ matrices	$2, 4, 6 \ldots 2l$	Self-representation
C_l	All symplectic $2l \times 2l$ matrices	$2, 4, 6 \ldots 2l$	Self-representation
D_l	All orthogonal $2l \times 2l$ matrices	$2, 4, 6 \ldots 2l-2, l$	Self-representation, one of the two fundamental spinor representations

j_i, $i = 1 \ldots l$, where l is the rank of the group. Hence, the invariants are functions of the $l j_i$. If, therefore, we could find a set of invariants such that, conversely, the j_i could be expressed in terms of these invariants, then all other invariants would depend on these (via the j_i) and the set would be complete. But since we have $l j_i$, we could expect to need just l of them. Furthermore, in each case they are the l lowest-order invariant we can form (for B_l, C_l, and D_l the odd-order invariants drop out). Hence it is (I hope!) quite plausible that these form the complete set we are looking for.

This brings us to the final problem (4). To express the invariants in a more tractable form, we go back to the form

$$\hat{I}_n = (\text{Sp } \hat{X}_\alpha \ldots \hat{X}_\gamma) X^\alpha \ldots X^\gamma \tag{17}$$

and note that this can also be written as

$$\hat{I}_n = \widehat{\text{Sp}}(\hat{X}_\alpha \ldots \hat{X}_\gamma) \times (X^\alpha \ldots X^\gamma) \tag{18}$$

where the $\wedge$ on the Sp means that it is to be taken only with respect to $\hat{X}_\alpha$ space. But then

$$\hat{I}_n = \widehat{\text{Sp}}(\hat{X}_\alpha \otimes X^\alpha)(\hat{X}_\beta \otimes X^\beta) \ldots (\hat{X}_\gamma \otimes X^\gamma) \tag{19}$$

$$= \widehat{\text{Sp}}(\hat{A})^n \tag{20}$$

where

$$\hat{A} = \hat{X}_\alpha \otimes X^\alpha \tag{21}$$

Equations (20) and (21) give the $\hat{I}_n$ in the required form, but to indicate what they really mean it is better to take an example. Let us go back to 0_3 and X_α as the self-representation. Then

$$\hat{X}_1 = \begin{pmatrix} 0 & 1 & 0 \\ -1 & 0 & 0 \\ 0 & 0 & 0 \end{pmatrix} \hat{X}_2 = \begin{pmatrix} 0 & 0 & -1 \\ 0 & 0 & 0 \\ 1 & 0 & 0 \end{pmatrix} \hat{X}_3 = \begin{pmatrix} 0 & 0 & 0 \\ 0 & 0 & 1 \\ 0 & -1 & 0 \end{pmatrix} \tag{22}$$

so that

$$\hat{A} = \begin{pmatrix} 0 & X_1 & -X_2 \\ -X_1 & 0 & X_3 \\ X_2 & -X_3 & 0 \end{pmatrix}. \tag{23}$$

Then, for example,

$$\hat{I}_2 = \text{Sp } \hat{A}^2 = \text{Sp} \begin{pmatrix} -X_1^2 - X_2^2 & & \\ & -X_1^2 - X_3^2 & \\ & & -X_2^2 - X_3^2 \end{pmatrix}$$

$$= -2(X_1^2 + X_2^2 + X_3^2) \tag{24}$$

Thus the idea is to put in the X_α as matrix elements in the $\hat{X}_\alpha$ representation, and then treat the resulting $\hat{A}$ as if its elements were numbers instead of matrices X_α. The method is dovetailed to suit the case where $\hat{X}_\alpha$ is a self-representation, which is just what we need according to Table I. (For the special "spinor invariant," it is easier to write it down directly.) We shall close by illustrating this for A_l:

For simplicity, we drop the unimodular condition, which is irrelevant for our purpose here; then for A_l the self-representation $\overset{s}{X}_\alpha$ consists of all the matrices $\overset{s}{X}_{ij} = |i\rangle\langle j|$ where the $|i\rangle$ is an orthonormal basis in the self-representation space, and we have switched from the single-index α to the double-index ij notation. Then, by definition

$$\hat{A} = |i\rangle\langle j|X_{ij} \tag{25}$$

so that $\hat{A}$ consists of the matrix with X_{ij} put exactly in the ij-th position. Thus

$$I_n = \text{Sp } \hat{A}_n = X_{ij}X_{jk}\ldots X_{lm}X_{mi} \tag{26}$$

which gives the $\hat{I}_n$ explicitly in terms of the $X_\alpha \equiv X_{ij}$ in a very simple way. This is the form of the I_n given by Okubo[4] for $SU_3 \subset A_2$.

REFERENCES

1. G. Racah, "Group Theory and Spectroscopy," CERN, Reprint 61-8 (1961).
2. G. Racah, *Rend. Lincei* **8**: 108 (1950).
3. B. Gruber and L. O'Raifeartaigh, *J. Math. Phys.* **5**: 1796 (1964).
4. S. Okubo, *Progr. Theoret. Phys.* **27**: 949 (1962).

On Peratization Methods[*]

N. R. RANGANATHAN

MATSCIENCE
Madras, India

---

1. INTRODUCTION

In this lecture we shall present the recent work of Pais and Feinberg[1-3] on weak interactions. This work is, in many respects, a bold and interesting step in the theory of weak interactions. Until now, weak-interaction phenomena were more or less analyzed using, to a large extent, arguments based on symmetry. The weak-interaction Lagrangian was only used "dynamically" to evaluate the second-order matrix elements. It was always believed that the effects due to higher order weak interactions could be neglected. Further, since the weak interactions are unrenormalizable (in the normal sense), no serious attempt was made to calculate the higher order graphs.

Weak interactions can be formulated either as four-fermion Fermi field theory or as the W-field theory, where we picture the weak interactions as being mediated by massive charged vector bosons (W). Though the W-theory is less singular, it is as unrenormalizable as the former. As we shall see, Pais and Feinberg have developed a technique of summing a class of graphs which yields a finite sum and which differs from the lowest-order calculation by a factor. This formal summation is equivalent to solving a modified integral equation. We shall not refer to the question of convergence of any iteration procedure that is discussed in this paper, since it is too early to discuss such questions.

---

[*] See the footnote in ref. 1 for the meaning of the word *peratization*.

2. HIGHER ORDER WEAK INTERACTIONS

In order to motivate the study of higher-order weak interactions, we shall demonstrate some of the possible effects. Hereafter, we shall discuss only the W-field theory, although at the end we shall show that, even though the W-field theory and four-fermion field theory can be made equivalent in the second order, this is not necessarily true when we sum all the higher order graphs.

By *higher order graph* we imply a graph in which a large number of vector mesons are exchanged between the fermions. Though it involves higher powers of g^2, we do not neglect them. As is to be expected, higher order weak interactions will also raise the problems of renormalization of mass and charge, but this can be discussed as in renormalizable theory. However, the most interesting effects are those on the selection rules which are only valid up to the second order in weak interactions.

Leptonic Reactions

Only leptons occur in the initial and final state. The Lagrangian in the two-component neutrino theory is

$$\mathscr{L}_{\text{lept}} = ig W_\rho [\bar{\mu}\gamma_\rho(1 + \gamma_5)\nu_\mu + \bar{e}\gamma_\rho(1 + \gamma_5)\nu_e] + \text{h.c.} \qquad (2.1)$$

An allowed process can occur with the exchange of a single vector

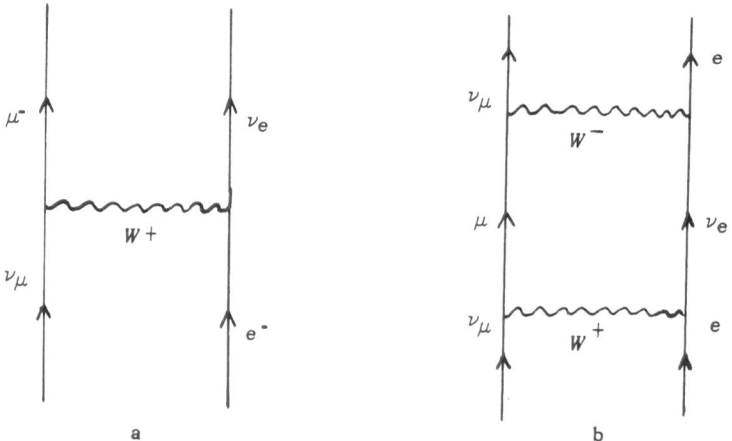

Fig. 1. (a) Allowed process. (b) Forbidden process.

meson, while the forbidden process needs the exchange of at least two vector mesons. We have, for example (Fig. 1),

$$\nu_\mu + e^- \longrightarrow \nu_e + \mu^- \qquad \text{(allowed)} \qquad (2.2a)$$

$$\nu_\mu + e^- \longrightarrow \nu_\mu + e^- \qquad \text{(forbidden)} \qquad (2.2b)$$

Semileptonic Reactions

These reactions involve leptons as well as strongly interacting particles. For the sake of illustration, consider as a model the interaction Lagrangian involving a charged vector meson and containing $\Delta S = 0$ and $\Delta S = \Delta Q = +1$ currents:

$$\mathscr{L}_{\text{int}} = ig_1 W_\rho \bar{n}\gamma_\rho(1 + \gamma_5)p + ig_2 W_\rho \bar{\Lambda}\gamma_\rho(1 + \gamma_5)p + \text{h.c.} + \dots \quad (2.3)$$

If we consider equations (2.1) and (2.3) together, it is easy to see that some of the allowed processes are

$$n \longrightarrow p + e^- + \bar{\nu}_e \qquad \Lambda \longrightarrow p + e^- + \bar{\nu}_e \qquad (2.4)$$

Again, we have forbidden processes, for example (Fig. 2),

$$\nu_e + p \longrightarrow \nu_e + p \qquad \text{(neutral lepton currents)} \qquad (2.5a)$$

$$\Sigma^+ \longrightarrow n + e^+ + \nu_e \qquad (\Delta S = 1, \Delta Q = -1) \qquad (2.5b)$$

$$\Xi^- \longrightarrow n + e^- + \bar{\nu}_e \qquad (\Delta S = 2) \qquad (2.5c)$$

Nonleptonic Processes

From the $K_1 - K_2$ mass difference which arises as a combined effect of strong and higher order weak interactions, we can conclude that the effect due to higher order weak interactions is smaller by a factor of 10^{-7} than the processes occurring in the second order. It is to be noted that four vector mesons need to be exchanged in order to cause a $K_1 - K_2$ mass difference. It is quite possible that for nonleptonic interactions, successive orders of g^2 are indeed smaller by factors of 10^{-7}. We can understand this in the case of nonleptonic interactions if we presume that there exist possible damping effects of strong interactions, because every vector meson must be emitted and absorbed by a strongly interacting particle. Finally, we note that with the inclusion of higher order weak interactions, lepton pairs need not emerge from a process at a single space-time point, and an experimental test of such nonlocal emission of lepton pairs may soon be available in the high-energy neutrino experiments.

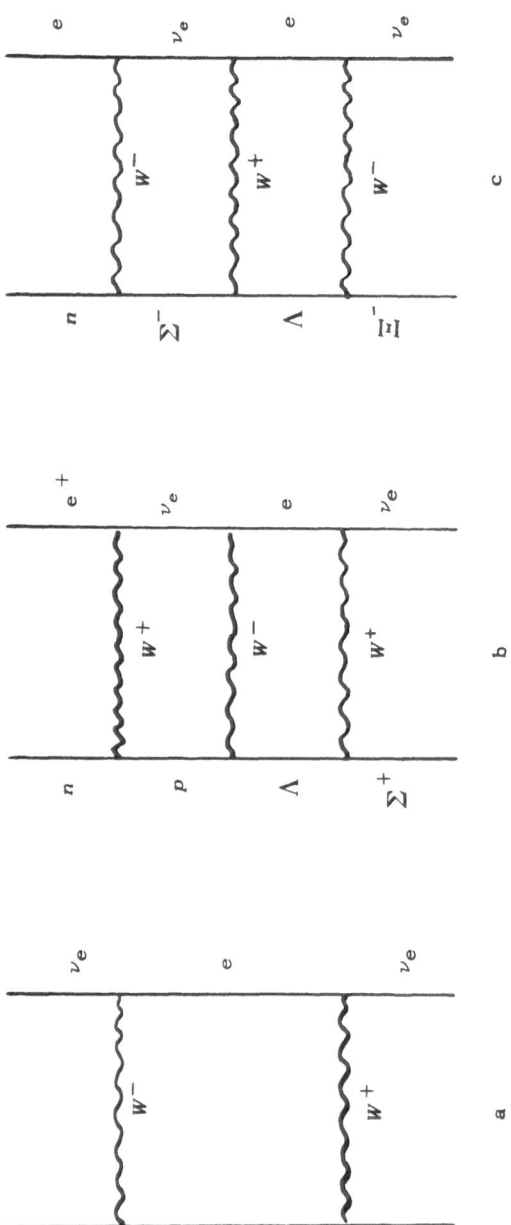

Fig. 2. Forbidden semileptonic processes. (a) Neutral lepton currents. (b) $\Delta S = \Delta Q$. (c) $\Delta S = -2$.

3. UNCROSSED LADDER GRAPHS

We will discuss the uncrossed ladder graphs for the process (Fig. 3)

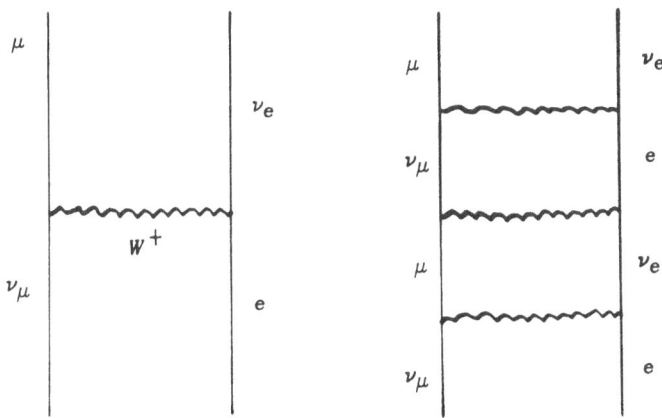

Fig. 3. Ladder graphs for $\nu_\mu + e \longrightarrow \nu_e + \mu$.

$$\nu_\mu + e \longrightarrow \nu_e + \mu$$

Since we wish to sum the leading singularities of each graph, we have to know the maximum degree of divergence which occurs in each order. To this purpose, we proceed as follows: Of course, we have to introduce a cutoff Λ for virtual momenta in order to estimate the degree of divergence. Consider a simple uncrossed ladder graph in which n vector mesons are exchanged. We note that:

1. Each integration over virtual momenta gives Λ^4.
2. Each fermion propagator introduces a factor Λ^{-1}, and there are two fermion propagators.
3. We shall simply take that the W-propagator counts as Λ^0.

This can be seen from the W-propagator, which is

$$-i\left[\frac{\delta_{\mu\nu} + m^{-2}q_\mu q_\nu}{q^2 + m^2}\right] \tag{3.1}$$

Only the second term need be considered for estimating maximum divergence. Collecting these, we find for a graph of order n that the leading singularity is proportional to

$$m^{-2}g^{2n+2}\left(\frac{\Lambda}{m}\right)^{2n} \qquad n = 0, 1, \ldots \tag{3.2}$$

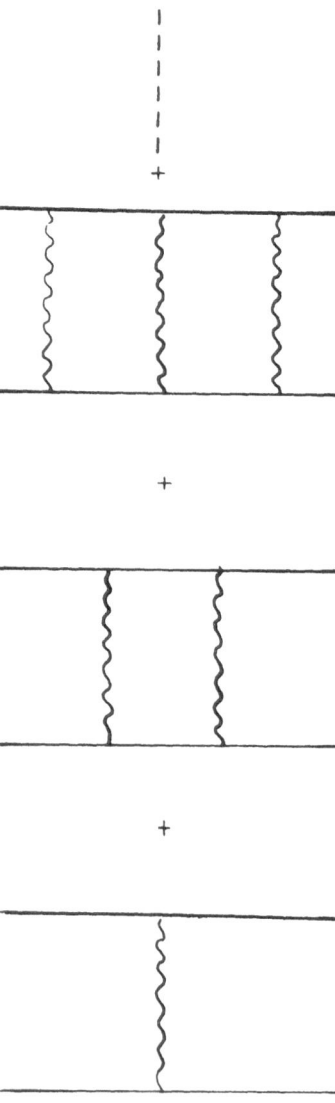

Fig. 4. Uncrossed ladder sum.

It is to be remembered that the allowed process takes place with the exchange of an odd number of vector mesons, while the forbidden process occurs with an even number. Let us sum over these singularities and write the sum as

$$g^2 F(x) \tag{3.3}$$

where

$$x = \left(\frac{g\Lambda}{m}\right)^2 \qquad F(x) = \sum_{n=0}^{\infty} \alpha_n x^n$$

The α_n do not depend on g or Λ. If we find that in the limit $\Lambda \to \infty$, $F(x)$ is finite, that is,

$$F(\infty) = \lim_{\Lambda \to \infty} F(x) = \text{finite} \tag{3.4}$$

our leading approximation is still $0(g^2)$. Thus, the higher order weak interactions may lead to observable effects in the lowest order.

We can similarly define the sum of the terms next to the leading singularities, which may again yield a finite sum. If we can continue this procedure still further, perhaps we may hope to have a new type of expansion. The above in essense constitutes the peratization program.

Let us now be more specific and see how it is done for a class of graphs. We shall consider the uncrossed ladder sum (Fig. 4). From equation (2.1) we have the following Feynman rules:

1. At each W-lepton vertex, insert a factor

$$g\gamma_\mu(1 + \gamma_5) \tag{3.5}$$

2. For each fermion propagator, insert a factor

$$S_F(p) = \frac{1}{\not{p}} \qquad \not{p} = i\gamma_\lambda p_\lambda \tag{3.6}$$

We neglect the lepton mass, which does not give rise to any difficulty. As is well known, we have to adopt a regularization procedure for the propagators in order that formal manipulations of them can be given a precise meaning. We can either regularize the fermion or meson propagator. We denote the regularized fermion propagator as $(1/\not{p})_R$, given by

$$\left(\frac{1}{\not{p}}\right)_R = -\not{p}\left(\frac{1}{p^2} - \frac{1}{p^2 + M^2}\right) \tag{3.7}$$

3. For each W-propagator, we insert a factor

$$\Delta_{\mu\nu}(q) = -i\left(\delta_{\mu\nu} + \frac{q_\mu q_\nu}{m^2}\right)\frac{1}{q^2 + m^2} \tag{3.8}$$

In this case, for the regularized propagator we take

$$[\Delta_{\mu\nu}(q)]_R = \left(\delta_{\mu\nu} + \frac{q_\mu q_\nu}{m^2}\right)\left(\sum_{i=1}^{M}\frac{\alpha_i}{q^2 + m_i^2}\right)$$

$$= \left(\delta_{\mu\nu} + \frac{q_\mu q_\nu}{m^2}\right)\langle(q^2 + m^2)^{-1}\rangle \tag{3.9}$$

Let us denote by $M^{(n)}$ the contribution of the ladder graph in which nW-mesons are exchanged.

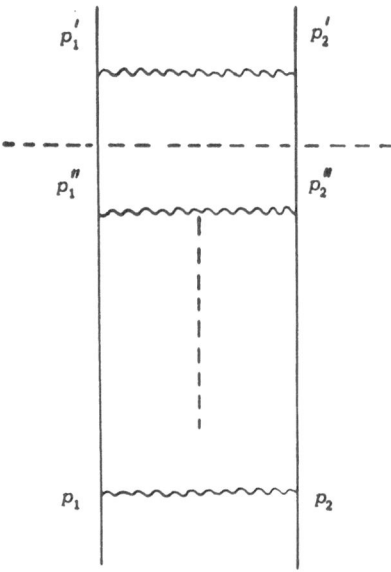

Fig. 5. The general uncrossed ladder graph.

In Fig. 5, p_1, p_2 denote the four momenta of the initial leptons and p_1', p_2' those of final leptons. We indicate the momentum transfers q and q' by

$$q = p_1' - p_1 = p_2 - p_2'$$
$$q' = p_1'' - p_1 = p_2 - p_2'' \tag{3.10}$$

We immediately write the equations for various $M^{(n)}$

$$M^{(1)}(q) = -ig^2\gamma_\mu^{(1)}(1 + \gamma_5^{(1)})\gamma_\nu^{(2)}(1 + \gamma_5^{(2)})$$
$$\times (\delta_{\mu\nu} + m^{-2}q_\mu q_\nu)\langle(q^2 + m^2)^{-1}\rangle \tag{3.11}$$

and

$$M^{(n+1)}(p_1', p_2'; p_1, p_2)$$
$$= \frac{-ig^2}{(2\pi)^4}\int \gamma_\mu^{(1)}(1 + \gamma_5^{(1)})\frac{1}{\not{p}_1''}\gamma_\nu^{(2)}(1 + \gamma_5^{(2)})\frac{1}{\not{p}_2''}M^{(n)}(p_1'', p_2''; p_1, p_2)$$
$$\times \left(\delta_{\mu\nu} + \frac{q_\mu q_\nu}{m^2}\right)\left\langle\frac{1}{q^2 + m^2}\right\rangle d^4 p_1'' d^4 p_2'' \delta^4(p_1'' + p_2'' - p_1 - p_2) \tag{3.12}$$

After we carry out integration over p_2'', we have $p_2'' = p_1 + p_2 - p_1''$. In equations (3.11) and (3.12), we have used the regularized W-meson propagator.

We can define the odd and even ladders as

$$M_{\text{odd}} = \lim_{N \to \infty} \sum_{n=0}^{N} M^{(2n+1)} \qquad \text{(allowed)}$$

$$M_{\text{even}} = \lim_{N \to \infty} \sum_{n=1}^{N} M^{(2n)} \qquad \text{(forbidden)}$$

We find that M_{odd} and M_{even} satisfy the following equations

$$
\begin{aligned}
M_{\text{odd}} = M^{(1)} - \frac{ig^2}{(2\pi)^4} \int \gamma_\mu^{(1)}(1 + \gamma_5^{(1)}) \frac{1}{\not{p}_1''} \gamma_\nu^{(2)}(1 + \gamma_5^{(2)}) \frac{1}{\not{p}_2''} \\
\times (\delta_{\mu\nu} + m^{-2} q_\mu q_\nu) \langle (q^2 + m^2)^{-1} \rangle M_{\text{even}} d^4 p_1''
\end{aligned}
\tag{3.14}
$$

$$
\begin{aligned}
M_{\text{even}} = -ig^2 \int \gamma_\mu^{(1)}(1 + \gamma_5^{(1)}) \frac{1}{\not{p}_1''} \gamma_\nu^{(2)}(1 + \gamma_5^{(2)}) \frac{1}{\not{p}_2''} \\
\times (\delta_{\mu\nu} + m^{-2} q_\mu q_\nu) \langle (q^2 + m^2)^{-1} \rangle M_{\text{odd}} d^4 p_1'' - R
\end{aligned}
\tag{3.15}
$$

where R is a remainder term which is assumed to vanish in the limit $N \longrightarrow \infty$. We now introduce new amplitudes $M_\pm$ by

$$M_\pm = M_{\text{odd}} \pm M_{\text{even}} \tag{3.16}$$

We find that the $M_\pm$ satisfy the equation

$$
\begin{aligned}
M_\pm = M^{(1)} \mp \frac{ig^2}{(2\pi)^4} \int \gamma_\mu^{(1)}(1 + \gamma_5^{(1)}) \frac{1}{\not{p}_1''} \gamma_\nu^{(2)}(1 + \gamma_5^{(2)}) \frac{1}{\not{p}_2''} \\
\times \left\langle \frac{1}{(p_1'' - p_1')^2 + m^2} \right\rangle \left\{ \delta_{\mu\nu} - \frac{(p_1'' - p_1')_\mu (p_2'' - p_2')_\nu}{m^2} \right\} \\
\times M_\pm d^4 p_1''
\end{aligned}
\tag{3.17}
$$

We now observe that the highest power of the integration variable p_1'' comes from $p_{1\mu}'' p_{2\nu}''$, and hence we introduce an approximation procedure by isolating this term; that is, we now write

$$M_\pm = M_\pm(0) + M_\pm(1) + \cdots \tag{3.18}$$

We expect $M_\pm(0)$ to be dominant. We note that on contacting $\gamma_\mu^{(1)}$ and $\gamma_\nu^{(2)}$ with $(p_1'')_\mu$ and $(p_2'')_\nu$, respectively, we cancel the fermion propagators. Thus, for $M_\pm(0)$ we have the equation

$$
\begin{aligned}
M_\pm(0) = M^{(1)} \pm \frac{ig^2}{m^2 (2\pi)^4} \\
\times \int (1 - \gamma_5^{(1)})(1 - \gamma_5^{(2)}) \left\langle \frac{1}{(p_1'' - p_1')^2 + m^2} \right\rangle M_\pm(0) d^4 p_1''
\end{aligned}
\tag{3.19}
$$

If we put

$$M_\pm(0) = \gamma_\mu^{(1)}(1 + \gamma_5^{(1)}) \gamma_\nu^{(2)}(1 + \gamma_5^{(2)}) M_{\mu\nu}^\pm \tag{3.20}$$

using equation (3.11) and linear independence of the γ_μ, we have

$$M_{\mu\nu}^\pm(q) = -ig^2\left(\delta_{\mu\nu} + \frac{q_\mu q_\nu}{m^2}\right)\left\langle\left(\frac{1}{q^3 + m^2}\right)\right\rangle$$
$$\pm \frac{4ig^2}{m^2(2\pi)^4}\int M_{\mu\nu}^\pm(q')\left\langle\frac{1}{(q'-q)^2 + m^2}\right\rangle d^4q' \tag{3.21}$$

Noting that the inhomogeneous term depends only on $q = p_1' - p_1$ and the kernel on $q' - q$, we see that $M_{\mu\nu}^\pm$ will depend on q, and hence

$$M_{\mu\nu}^\pm(q) = -ig^2\left(\delta_{\mu\nu} + \frac{q_\mu q_\nu}{m^2}\right)\left\langle\frac{1}{q^2 + m^2}\right\rangle$$
$$\pm \frac{4ig^2}{m^2(2\pi)^4}\int M_{\mu\nu}^\pm(q')\left\langle\frac{1}{(q'-q)^2 + m^2}\right\rangle d^4q' \tag{3.22}$$

The above integral equation implies that all the W-mesons are emitted at one point and absorbed at another point. We call the graphs corresponding to equation (3.22) collapsed ladder graphs (Fig. 6).

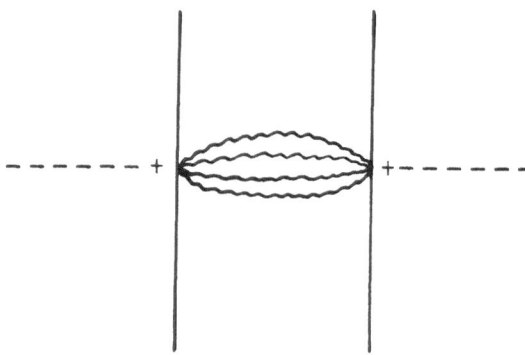

Fig. 6. Collapsed ladder graph.

If we define the Fourier transform of $M_{\mu\nu}^\pm(q)$ in coordinate space as $M_{\mu\nu}^\pm(y)$

$$M_{\mu\nu}^\pm(y) = \int M_{\mu\nu}^\pm(q)\frac{e^{-iqy}}{(2\pi)^4}d^4q \tag{3.23}$$

we have

$$M_{\mu\nu}^\pm(y) = \frac{-ig^2[\delta_{\mu\nu} - (\partial_\mu \partial_\nu/m^2)]\bar{\Delta}_F(y)}{1 \pm [(-4ig^2/m^2)\bar{\Delta}_F(y)]} \tag{3.24}$$

where $\bar{\Delta}_F(y)$ is the regularized Feynman propagator in coordinate space. The above solution is still purely formal since we have yet to carry out both integrations over singular functions and the limiting process $M \longrightarrow \infty$ as a part of the regularization procedure. These two are fairly complicated because they involve detailed considerations of integrals over various Bessel functions, which have been done by Feinberg and Pais.[1]

In order to see what new result we obtain from equation (3.24), we now write

$$M^{\pm}_{\mu\nu}(q) = \alpha^{\pm}(q^2)\delta_{\mu\nu} + \beta^{\pm}(q^2)q_{\mu}q_{\nu} \qquad (3.25)$$

Since q is the only vector, equation (3.25) can be written. Using equation (3.25) and the Fourier relationship between $M^{\pm}_{\mu\nu}(q)$ and $M^{\pm}_{\mu\nu}(y)$, we can write for the trace of (3.25) the equation

$$4\alpha^{\pm} + q^2\beta^{\pm} = -ig^2 \int \frac{e^{iqy}(4 - m^{-2}\square)\Delta_F(y)}{1 \pm [(-4ig^2/m^2)\Delta_F(y)]} d^4y \qquad (3.26)$$

since $\square\Delta_F(y) = m^2\Delta_F(y) - \delta^4(y)$,

$$4\alpha^{\pm} + q^2\beta^{\pm} = -ig^2 \int \frac{e^{iqy}[3\Delta_F(y) + m^{-2}\delta^4(y)]}{1 \pm [(-4ig^2/m^2)\Delta_F(y)]} d^4y \qquad (3.27)$$

As $\Delta_F(0) = \infty$, we note the integral involving $\delta^4(y)$ vanishes. We see that this vanishing comes about only after summing all the diagrams, which causes the second term in the denominator. If we had taken only the second-order diagram, this damping of singularity would not have happened. Finally,

$$4\alpha^{\pm}(q^2) + q^2\beta^{\pm}(q^2) = -3ig^2 \int \frac{e^{iqy}\Delta_F(y)d^4y}{1 \pm [(-4ig^2/m^2)\Delta_F(y)]} \qquad (3.28)$$

If we assume $\beta^{\pm}(q^2)$ is well behaved at $q^2 = 0$, then for $q^2 = 0$, i.e., at zero energy, we have

$$\alpha^{\pm}(q^2 = 0) = \frac{-3ig^2}{4} \lim_{q^2 \to 0} \int \frac{e^{iqy}\Delta_F(y)d^4y}{1 \pm [(-4ig^2/m^2)\Delta_F(y)]} \qquad (3.29)$$

Pais and Feinberg have carried out the regularization procedure and have found the integral in the right-hand side of equation (3.29) to be

$$\frac{1}{q^2 + m^2} \pm 0(g^2 \ln g)$$

Hence

$$\alpha^{\pm}(q^2 = 0) = \frac{-3ig^2}{4m^2} \pm 0(g^4 \ln g) \qquad (3.30)$$

From this, we deduce

$$\alpha_{\text{allowed}}(q^2 = 0) = \frac{-3ig^2}{4m^2}$$

$$\alpha_{\text{forbidden}}(q^2 = 0) = 0(g^4 \ln g) \qquad (3.31)$$

For the lowest-order graph alone, we have

$$\alpha_{\text{allowed}}(q^2 = 0) = \frac{-ig^2}{m^2} \qquad \alpha_{\text{forbidden}}(q^2 = 0)$$

We thus see that the effect of higher order weak interactions is that the lowest order amplitude is reduced by a factor $\frac{3}{4}$. However, this $\frac{3}{4}$ can be absorbed by redefining g^2, but the other physical quantities will reflect this redefinition of g^2. In case of μ decay, while in the peratization theory Michael parameter is given by $(\frac{3}{4}) + (\frac{4}{9})(m_\mu/m)^2$, we have in perturbation theory $\frac{3}{4} + \frac{1}{3}(m_\mu/m)^2$. These effects are very small to be detected at present.

4. TRACE EQUATION

The procedure outlined in the preceding section is not very convenient for iteration. We now discuss an alternative procedure in which we treat the full amplitude instead of $M_\pm(0)$ as before. Since we are interested in the leading correction to g^2, this can be obtained by setting $q^2 = 0$. We now study the full amplitude after setting from the beginning $p_1 = p_2 = 0$. Now the amplitude depends only on one four-momentum. Denoting the full amplitude $M_{\mu\nu}^\pm(p, M)$ where $p = p_1' = -p_2'$, we can set

$$M_{\mu\nu}^\pm(p, M) = \alpha^\pm(p, M) \pm \beta^\pm(p, M)p_\mu p_\nu \qquad (4.1)$$

We now find that the trace of $M_{\mu\nu}^\pm(p, M)$, that is, $T^\pm(p, M)$, satisfies

$$
\begin{aligned}
T^\pm(p, M) = &\frac{-ig^2[4 + (p^2/m^2)]}{p^2 + m^2} \pm \frac{4ig^2 M^4}{(2\pi)^4} \\
&\times \int \frac{d^4 p'}{p'^2(p'^2 + M^2)^2[(p' - p)^2 + m^2]}\left[4 + \frac{(p' - p)^2}{m^2}\right] T^\pm(p, M)
\end{aligned} \qquad (4.2)
$$

where

$$T^\pm(p, M) = 4\alpha^\pm(p, M) \pm p^2 \beta^\pm(p, M) \qquad (4.3)$$

Using equations (4.2) and (4.3), we can obtain an equation, for Pais and Feinberg have shown that

$$\lim_{M \to \infty} \lim_{p \to 0} p^2 \beta^\pm(p, M) = 0$$

Hence

$$\alpha^\pm = \lim_{M \to \infty} \lim_{p \to 0} \alpha^\pm(p, M) = \tfrac{1}{4} T^\pm \qquad (4.4)$$

where

$$T^{\pm} = \lim_{M \to \infty} \lim_{p \to 0} T^{\pm}(p, M) \qquad (4.5)$$

The $p \to 0$ limit is necessary because $M^{\pm}_{\mu\nu}(p, M)$ is actually off-shell, and only $p = 0$ gives the physically meaningful on-the-shell amplitude. Thus, at zero energy

$$M^{\pm} = \alpha^{\pm} \gamma^{(1)}_{\mu} (1 + \gamma^{(1)}_5) \gamma^{(2)}_{\mu} (1 + \gamma^{(2)}_5) \qquad (4.6)$$

for solving (4.2) we put

$$T^{\pm}(p, M) = T^{\pm}_1(p, M) + T^{\pm}_2(p, M) \qquad (4.7)$$

where

$$T^{\pm}_1(p, M) = \frac{-ig^2}{m^2} \pm \frac{4ig^2 M^4}{(2\pi)^4 m^2} \int d^4 p_1 \left\{ \frac{T^{\pm}_1(p', M) + T^{\pm}_2(p', M)}{p'^2 (p'^2 + M^2)^2} \right\} \qquad (4.8)$$

$$T^{\pm}_2(p, M) = \frac{-3ig^2}{p^2 + m^2}$$
$$\pm \frac{12ig^2 M^4}{(2\pi)^4} \int \frac{d^4 p' \, T^{\pm}_1(p', M)}{p'^2 (p'^2 + M^2)^2 [(p - p')^2 + m^2]} \qquad (4.9)$$
$$\pm \frac{12ig^2 M^4}{(2\pi)^4} \int \frac{d^4 p' \, T^{\pm}_2(p', M)}{p'^2 (p'^2 + M^2)^2 [(p - p')^2 + m^2]}$$

We note that from equation (4.8) $T^{\pm}_1(p, M)$ is actually independent of p. Thus, it is proper to call $T^{\pm}_1$ the high-frequency part, since it will contribute even at high virtual frequencies. $T^{\pm}_2(p, M)$ is called the low-frequency part.

In order to solve equations (4.8) and (4.9), we first compute the "Born approximation" from equation (4.9) for $T^{\pm}_2(p, M)$. We find

$$T^{\pm}_2(0, M) = \frac{-3ig^2}{m^2} + \frac{3g^2 T^{\pm}_1(M)}{(2\pi)^2} \ln \frac{M}{m} \qquad (4.10)$$

In order to study the behavior of $T^{\pm}_2(0, M)$ as $M \to \infty$, we now solve equation (4.8) after substituting for $T^{\pm}_2(p, M)$ expressions of order g^2. Now we find that the calculated $T^{\pm}_1(M)$ vanishes as $M \to \infty$. Thus, to the leading order we have

$$T^{\pm}_2(0, M) = \frac{-3ig^2}{4m^2} \qquad (4.11)$$

Hence, we find

$$M^{\pm}_{\mu\nu} = \frac{-3ig^2}{4m^2} \delta_{\mu\nu} \qquad (4.12)$$

which implies that

$$M_{\mu\nu,\text{odd}} = \frac{-3ig^2}{4m^2} \delta_{\mu\nu} \qquad (4.13)$$

and

$$M_{\mu\nu,\text{even}} = 0 \tag{4.14}$$

Equations (4.13) and (4.14) coincide with equation (3.31).

Pais and Feinberg evaluated the second Born approximation for $T_{\frac{1}{2}}^{\pm}(p, M)$, and they found the leading term for the forbidden process to be

$$M_{\mu\nu,\text{even}} = \frac{9ig^4}{16\pi^2 m^2}\delta_{\mu\nu} \tag{4.15}$$

We shall briefly remark on the comparison between the four-fermion local coupling theory and the W-meson theory. It is known that in the lowest-order the former can be considered as the limit of the W-theory for infinite boson mass. In particular, we let $g \longrightarrow \infty$, $m \longrightarrow \infty$ such that

$$\frac{g^2}{m^2} = \frac{G}{\sqrt{2}} \tag{4.16}$$

where G is the Fermi coupling constant. The question arises as to what will happen if we sum the graphs in the four-fermion theory which correspond to ladder graphs in W-meson theory. The graphs in the Fermi theory are obtained by shrinking the W-rungs (Fig. 7). Using a procedure almost identical to that in W-theory to evaluate the sum of such graphs, we now find, however, that the amplitude damps itself

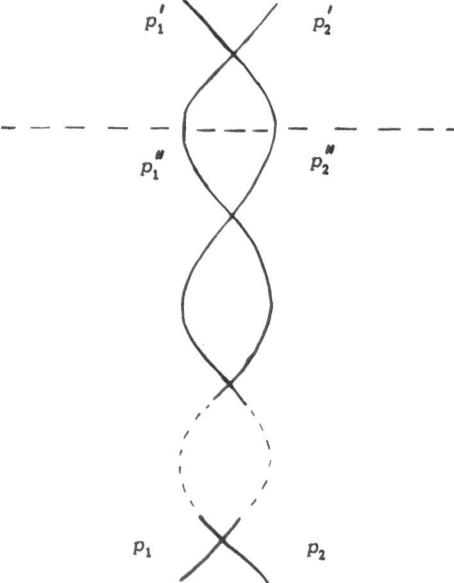

Fig. 7. "Ladder" graphs in the Fermi theory.

out, in contrast to the case in W-meson theory, where we obtain a finite result. It is not clear whether the inclusion of a larger class of graphs will make the Fermi theory nonvanishing.

5. CONCLUDING REMARKS

Throughout the above discussion we have tacitly assumed that the ladder graphs are the leading ones, i.e., the most divergent graphs. Some more recent work seems to indicate that this may not be so. Wu and Pwu[4] noted that W-W scattering through the leptonic loop is itself divergent and is of the order $g^4 \ln (\Lambda/m)$. We can now envisage a ladder in which the intermediate W-mesons scatter each other before being absorbed, i.e., *a dotted graph* (Fig. 8).

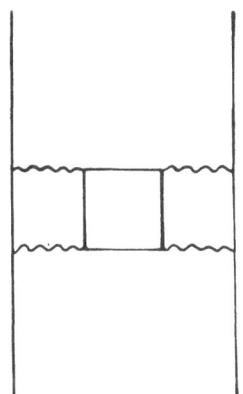

Fig. 8. Ladder with W-W scattering.

Since such scatterings can happen any number of times, we can sum the various dotted graphs. In fact, the prescription now will be to sum all the graphs which have the maximum number of dots to a given order of g. Wu and Pwu obtain a factor of $\frac{3}{8}$ instead of $\frac{3}{4}$. Now it is not enough to consider just the W-W scattering in the lowest order, i.e., the square diagram. Actually, instead of renormalizing the W-W scattering, we should peratize W-W scattering, i.e., we find the sum of the series. (Fig. 9). We should insert such a sum for each dot in the ladder graphs and, in fact, we now obtain an expansion

$$\frac{g^2}{m^2} \left[1 + \frac{a_2}{\ln g^2} + \frac{a_3}{(\ln g^2)^{3/2}} + \cdots \right] + 0(g^4)$$

The above remarks make it abundantly clear that the question of

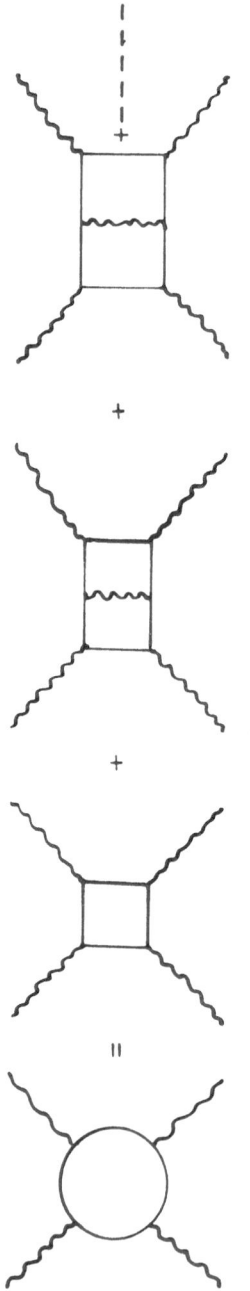

Fig. 9. *W-W* scattering.

leading graphs is still to be decided. It is quite possible that we may have to resort to nondiagrammatic methods for further progress in the peratization program.

REFERENCES

1. G. Feinberg and A. Pais, *Phys. Rev.* **131**: 2724 (1963).
2. G. Feinberg and A. Pais, *Phys. Rev.* **133**: B477 (1964).
3. A. Pais, "Methods and Problems in the Dynamics of Weak Interactions," Report in the 1963 SIENNA International Conference on Elementary Particles (preprint).
4. Y. Pwu and T.T. Wu, "Higher Order Leptonic Weak Interactions," *Phys. Rev.* **133B**: 1299 (1284).

See also:

M.A. Beg, *Ann. Phys.* **27**: 183 (1964).

H.M. Fried, *Phys. Rev.* **133B**: 1562 (1964).

W. Kummer, *Nuovo Cimento* **32**: 1653 (1964).

J.C. Pockinghorne and J.C. Taylor, *Phys. Rev.* **134B**: 420 (1964).

Large-Angle Elastic Scattering at High Energies*

R. HAGEDORN

CERN
Geneva, Switzerland

1. INTRODUCTION

High-energy scattering under reproducible conditions has been investigated for primary proton energies of up to 30 GeV at Brookhaven and CERN. The process can be divided into three main types:

1. *Inelastic scattering.* Very little is known theoretically, and the peripheral as well as the statistical model does not work very satisfactorily. Although rough estimates of production rates and spectra can be made, we are far from having any knowledge about the relevant amplitudes.

2. *Small-momentum-transfer elastic scattering.* Here, recent work on amplitudes, starting with the Mandelstam representation, Regge poles, and the multiperipheral model,[1] has at least yielded definite predictions which can be compared to experiments. One of these is that the elastic differential cross section should behave as

$$\frac{d\sigma_{el}}{dt} = \left(\frac{d\sigma_{el}}{dt}\right)_{t=0} \cdot F(t)\left(\frac{s}{2m^2}\right)^{2[\alpha(t)-1]} \tag{1}$$

for fixed t (momentum transfer squared) and s (center of momentum energy squared) going to infinity.

3. *Large-angle elastic scattering.* Here again, very little is known. The approximations to the Mandelstam representation, peripheral and multiperipheral models, and the Regge pole hypothesis all fail. It seems, however, that the statistical model describes the situation

---

* A final version of this contribution with generalization to processes like $p + p \rightarrow A + B$ and $\pi + p \rightarrow A + B$ and containing a comparison with the experiment is being published in *Nuovo Cimento*.

fairly well. The reason is that large-angle elastic scattering is related to small impact parameters, i.e., central collisions, in which practically all center of momentum energy becomes available for particle production—the very condition for establishing thermodynamic equilibrium.

We shall in the following discuss these points in detail:

First, the derivation of formula (1) will be sketched and its invalidity for large angles be shown; second, the statistical model will be applied to the large-angle elastic scattering and its application be justified; third, the results will be compared to recent experiments.

2. REGGE POLE ANALYSIS OF THE DIFFRACTION REGION AND ITS INVALIDITY FOR LARGE-ANGLE SCATTERING

Notation

We picture the scattering process—including pp-scattering, but ignoring spin and isospin—as shown in Fig. 1, where p, k, p', k' are

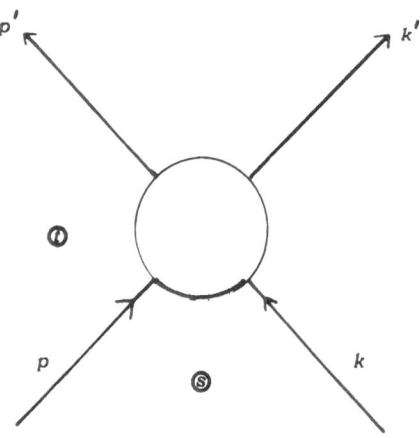

Fig. 1. Notation.

four vectors. If we define the scattering amplitude A via the S-matrix by

$$\langle f | S | i \rangle = \delta_{fi} + \frac{i}{(2\pi)^2} \frac{\delta^4(p + k - p' - k')}{(16 p_0 k_0 p_0' k_0')^{1/2}} A(pkp'k') \qquad (2)$$

then $A(pkp'k')$ is a relativistic invariant. It can depend only on two of the ten invariants which can be formed with the four four-vectors, since $p^2 = k^2 = p'^2 = k'^2 = m^2$ is uninteresting and the remaining six invariants are reduced to two by the four-equations $p + k - p' - k' = 0$.

We choose the two invariants s and t:

$$s = (p + K)^2 = 4(K_s^2 + m^2) = -2K_t^2(1 - \cos \theta_t)$$
$$t = (p - p')^2 = -2K_s^2(1 - \cos \theta_s) = 4(K_t^2 + m^2)$$
(3)

The "s-channel" is given for $s \geqslant 4m^2$, $t \leqslant 0$ and represents pp-scattering. The "t-channel" is given for $t \geqslant 4m^2$, $s \leqslant 0$ and represents $p\bar{p}$-scattering.

Accordingly,

K_s and θ_s are the (center of momentum) magnitude of the three-momentum and the scattering angle of either particle in the s-channel. (4)

K_t and θ_t are the corresponding quantities in the t-channel.

These quantities have physical values only in their respective channels, but, once defined, they will also be used if they have unphysical values.

If we calculate in the usual way the differential elastic cross section and—by means of the optical theorem—the total cross section, then we obtain in the present notation in the s-channel

$$\frac{d\sigma_{el}}{dt} = \frac{\pi}{K_s^2} \frac{d\sigma_{el}}{d\Omega} = \frac{|A(s,t)|^2}{|6\pi K_s^2 s|}$$

$$\sigma_{el} = \int \frac{d\sigma_{el}}{dt} dt = \frac{1}{64\pi K_s^2 s} \int |A(s,t)|^2 \, dt$$
(5)

$$\sigma_{tot} = \frac{1}{2K_s(s)^{1/2}} \, \text{Im} \, A(s,0)$$

Here, $A(s, t)$ is the invariant scattering amplitude.

Partial Wave Expansion

The important point now is that $A(s, t)$ is an analytic function of s and t and describes pp-scattering for $s \geqslant 4m^2$, $t \leqslant 0$ and $p\bar{p}$-scattering for $s \leqslant 0$, $t \geqslant 4m^2$. Thus it is one and the same function in all channels; the channels themselves merely represent particular ranges of the variables. If we now write a partial wave expansion in $P_l(\cos \theta)$ in either the s- or the t-channel, then both must represent the same analytical function $A(s, t)$:

$$A(s, t) = -4\pi i \frac{s^{1/2}}{K_s} \sum (2l + 1) P_l (\cos \theta_s) a_l^{(s)}(s)$$

in the s-channel

(6)

$$A(s, t) = -4\pi i \frac{t^{1/2}}{K_s} \sum (2l + 1) P_l (\cos \theta_t) a_l^{(t)}(t)$$

in the t-channel

Although both expansions represent the same function, we cannot conclude that the upper formula remains valid if s becomes negative and $t \geqslant 4\,m^2$—that is, if continued from the s- to the t-channel (and vice versa)—the reason is that under such a continuation the expansions will not converge.

Regge's Representation

Regge has shown how to overcome this difficulty and find a representation which can be continued from one channel to the other.[2] Unfortunately, his proof applies to potential scattering, where only one channel exists. Taking for granted the hypothesis that the result may nevertheless be applied to our case, we obtain a limiting formula (1) for high-energy elastic scattering in the following way:

We introduce $\lambda = l + il_1$, as complex angular momentum and write, in the t-channel,

$$A(s, t) = \frac{2\pi(t)^{1/2}}{K_t} \int_{C_0} \frac{2\lambda + 1}{\sin \pi \lambda} P_\lambda(-\cos \theta_t) a_\lambda^{(t)}(t)\, d\lambda \qquad (7)$$

where C_0 is the contour displayed in Fig. 2a. By Regge's assumptions (superposition of Yukawa potentials), it can be proved that the $a_l^{(t)}(t)$

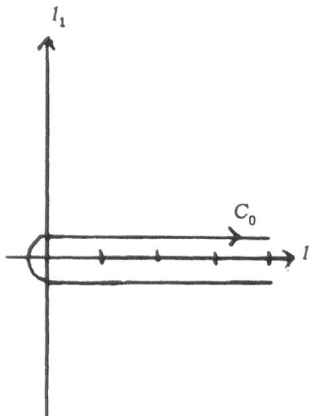

Fig. 2a. Contour C_0.

can be continued to $l \rightarrow \lambda$ and that equation (7) is then a consequence of Cauchy's formula. Moreover, the function $a_l^{(t)}(t)$ of λ has only a finite number of poles ("Regge poles") to the right of $\operatorname{Re} \lambda = -\frac{1}{2}$ and these lie in the upper half-plane. Finally, the contour C_0 may be deformed according to Fig. 2b; the infinite half-circles and the straight

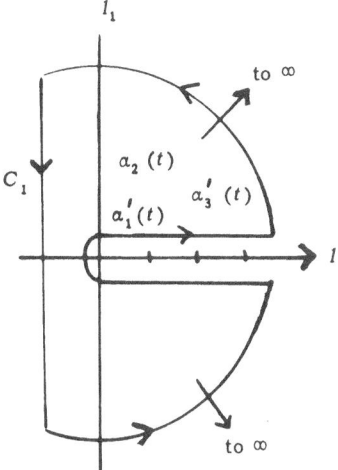

Fig. 2b. Contour C_1.

section from $+i\infty - \frac{1}{2}$ to $-i\infty - \frac{1}{2}$ vanish (the latter only for t fixed and $s \rightarrow \infty$, which we shall indeed obtain). Hence,

$$\int_{C_0} = \int_{C_1} = 2\pi i \, [\text{sum of residues of } a_\lambda^{(t)}(t)] \tag{8}$$

Let $\lambda_j \equiv \alpha_j(t)$ be the locations of the poles of $a_\lambda^{(t)}(t)$ and $R_{\alpha_j(t)}(t)$ the residues. Then we obtain from equations (7) and (8)

$$
\begin{aligned}
A(s,t) &= \frac{2\pi(t)^{1/2}}{K_t} \int_{C_0} \frac{2\lambda+1}{\sin \pi\lambda} P_\lambda(-\cos \theta_t) a_\lambda^{(t)}(t) d\lambda \\
&= \frac{2\pi(t)^{1/2}}{K_t} 2\pi i \sum_{j=1}^{N} \frac{2\alpha_j+1}{\sin \pi\alpha_j} R_{\alpha_j} P_{\alpha_j}(-\cos \theta_t)
\end{aligned} \tag{9}
$$

The position $\alpha_j(t)$ of the Regge poles is, of course, a function of t, as well as the values $R_{\alpha_j}(t)$.

Now comes the point: this finite sum can be continued from $t \geqslant 4m^2$, $s \leqslant 0$—for which it was written—to $t \leqslant 0$, $s \geqslant 4m^2$, since no convergence problem arises. Supposing then that the argument could be carried over to high-energy scattering (where the supposition of a potential no longer holds), then the region $t \leqslant 0$, $s \geqslant 4m^2$, which in potential scattering is highly unphysical, becomes physical again and describes elastic scattering in the s-channel.

Letting $s \longrightarrow \infty$ and keeping $t \leqslant 0$ fixed, we have from equation (3)

$$K_t = \left(\frac{t}{4} - m^2\right)^{1/2}$$

$$-\cos\theta_t = -1 - \frac{2s}{t - 4m^2} \longrightarrow \frac{2s}{|t| + 4m^2} \longrightarrow \infty$$

(10)

Since $P_\alpha(x) \longrightarrow x^\alpha$ if $x \longrightarrow \infty$ in equation (9), only the contribution from the $\alpha_j(t)$ with the largest real part of $\alpha_j(t)$ will survive for $s \longrightarrow \infty$. If we call the location of the pole lying farthest to the right $\alpha(t) \equiv \alpha_{\max}(t) + i\beta(t)$, equation (9) gives

$$\lim_{\substack{s \to \infty \\ t \text{ fixed}}} A(s, t) = g(t)\left(\frac{s}{2m^2}\right)^{\alpha(t) + i\beta(t)}$$

(11)

where all t-dependence has been lumped together in $g(t)$. Accordingly, with equation (5), the differential cross section becomes

$$\left.\frac{d\sigma_{\text{el}}}{dt}\right|_{\substack{s \text{ large} \\ t \text{ fixed}}} = \frac{1}{64\pi s K_s^2}|A(s, t)|^2$$

$$\longrightarrow \frac{|g(t)|^2}{16\pi s^2}\left(\frac{s}{2m^2}\right)^\alpha \equiv G(t)\left(\frac{s}{2m^2}\right)^{2[\alpha(t)-1]}$$

(12)

which may be rewritten in the form of equation (1).

A formula of this type has been experimentally confirmed for pp-scattering at high energies, although the implied shrinkage with increasing s of the diffraction peak seems to be absent in πp-scattering[3]; probably, at the now available energies, the lower poles of equation (9) still contribute appreciably. Whereas this theory may be adequate for fixed t and $s \longrightarrow \infty$, we see immediately that this is not so for fixed scattering angle $\cos\theta_s = 1 + (2t)/(s - 4m^2)$: If s goes to infinity and $\cos\theta_s$ is to be constant, t has to become proportional to s. In particular, if we consider large-angle elastic scattering, t will become of the order of s. Let us take $\theta_s = 90°$; then for $s \longrightarrow \infty$, $t \longrightarrow -(\frac{1}{2})s$, and consequently, in equation (10)

$$-\cos\theta_t = -1 - \frac{2s}{t - 4m^2} \longrightarrow 1 - \frac{2s}{-\frac{1}{2}s} = +3 = \text{constant}$$

Feeding that into the sum over Regge-residua (9) will, of course, no longer cause one or a few terms to be dominant; hence formula (12) certainly does not apply to large fixed-angle elastic scattering.

3. STATISTICAL MODEL APPROACH

It is for these large angles that the statistical model offers a way of calculating the differential cross section, if some additional reasonable assumptions are made. To make clear what the model means and what the additional assumptions are, we shall discuss both shortly.

The Essential Idea of the Statistical Model

The sketch given here is a brief resume of the article by the author, which may be consulted for details.[4]

The probability for any of the many possible channels (labeled by $j = 0, 1, 2, \ldots ; 0 \equiv$ elastic channel)

Channel label

$$p + p \rightarrow p + p \qquad\qquad 0$$

$$p + p + \pi^0 \qquad\qquad 1$$

$$n + p + \pi^+ \qquad\qquad 2$$

. .

. .

. .

$$N + N + n\pi \qquad\qquad .$$

$$N + Y + K \qquad\qquad .$$

. j

. .

$$N + Y + K + n\pi \qquad\qquad .$$

$$N + N + \bar{N} \ldots \qquad\qquad .$$

is given, up to a common constant, by the expression, valid in the center of momentum frame,

$$P_j = \int |\langle f_j | S | i \rangle|^2 \, d^3 p_1 \ldots d^3 p_n$$

$$= \prod_{i=1}^{n} \Omega_i f(T, \tau_1 \ldots \tau_n) \int \underbrace{d^3 p_1 \ldots d^3 p_n \delta(\sum \vec{p}_i) \, \delta(E - \sum E_i)}_{\text{phase space factor}} \tag{13}$$

Here use has been made of four-momentum conservation; $p_1 \ldots p_n$

are the momenta of the n-particles produced in channel j; $f(T, \tau_1 \ldots \tau_n)$ takes into account isospin and statistics, whereas the product of the "interaction volumes" Ω_i is the average of the reduced matrix element squared with respect to the integration over the phase space. From arguments considering the inverse process, Ω_i should be of the order of $(1/2\pi)^3$ times the nucleon volume for nucleons, pions, etc., and somewhat smaller for strange particles. Taking them independent of the primary energy means shifting everything to the phase-space factor and, in fact, is equivalent to the assumption that all center of momentum energy E is available for the production of particles. This is certainly not true and has to be corrected for: How to do this is one of the additional assumptions.

What, then, do we expect from the statistical model? If the energy E in the center of momentum system becomes large, then not only does the number of open channels grow rapidly, but also the values P_j themselves increase, and those of the many-particle channels grow faster than P_0, the probability for the elastic channels.

If now, roughly speaking, P_0 and $\sum P_j$ are proportional to the non-diffractive elastic and total inelastic cross sections, respectively, then we expect that the ratio $\sigma_{el}/\sigma_{inel}$ is a rapidly decreasing function of the total energy—simply because the one elastic channel cannot successfully compete with the many and strong inelastic ones.

Application of the Statistical Model

In the above form, equation (13) is reasonable only in central collisions, that is, in such collisions where the impact of the particles is so close that the whole of E is distributed in a "thermal equilibrium." These collisions will form only a part of all collisions. A rough estimate is based on the consideration that the duration t of impact should be long enough to allow the equilibrium to spread out over the whole of the two nucleons (see Fig. 3). Since in the center of momentum system the nucleons have elliptical shape, with "thickness"

$$a \approx \frac{2r}{\gamma}$$

the time of contact will be roughly $t \approx a \approx (2r/\gamma)$ and during this time (i.e., before the nucleons can part from each other) the interaction must not only spread over the whole distance ρ but even allow some thermodynamic equilibrium to be established. Hence, ρ must be so small that during t the interaction may go to and fro a few times;

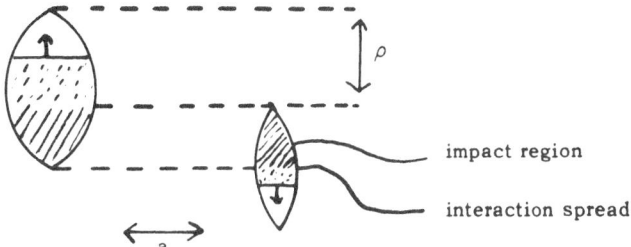

Fig. 3. The "central collision."

assuming at least two times and putting the velocity of propagation equal to $c = 1$, we see that $\rho \lesssim t/2 = r/\gamma$ is necessary for the impact parameter ρ. Since $\gamma = E/2m$, we find that the condition for "central collision" is $\rho \leqslant 2rm/E$, and we expect therefore

$$b = \frac{\text{Number of central collisions}}{\text{Number of all collisions}} \lesssim \frac{\pi\rho^2}{\pi r^2} \leqslant \frac{4m^2}{E^2} \qquad (14)$$

For a primary proton momentum of 25 GeV/c, we would obtain for equation (14) the numerical value 0.07. This agrees well with the experimental finding that for this primary momentum the number of antinucleons produced is roughly ten times smaller than that calculated with the statistical model, if the latter is applied without correcting for the fact that only approximately 0.1 of all collisions are central and make the total energy available. (The antinucleon production rate is very sensitive to the available energy, since at 25 GeV/c, the center of momentum energy is 7.6 proton masses—not very far from the threshold, which is four proton masses.)

We certainly must require rather central collisions for large-angle scattering and shall adopt the *assumption* that equation (14) gives the ratio correctly.

Large-Angle Elastic Scattering Dealt with by the Statistical Model[5]

Let us define the "compound elastic cross section"

$$\sigma_{c,\text{el}} \equiv \sigma_{\text{el}} - \sigma_{\text{diffr}} \qquad (15)$$

which is actually what we propose to calculate. Similarly, we define the nondiffractive part of the total cross section

$$\sigma_{c,\text{tot}} \equiv \sigma_{\text{tot}} - \sigma_{\text{diffr}} \approx \sigma_{\text{inel}} \qquad (16)$$

In these formulas σ_{diffr} is defined by the next additional *assumption*, namely, that the compound elastic differential cross section $d\sigma_{c,el}/d\Omega$ is isotropic, i.e.,

$$\frac{d\sigma_{c,el}}{d\Omega} = \frac{1}{4\pi}\sigma_{c,el} \tag{17}$$

This is rather plausible, because of the following mechanism: The two nucleons fuse into a compound system, in which a kind of thermodynamic equilibrium is established, and then, by pure chance, choose the elastic channel among all the many possible ones. Meanwhile, they have lost all memory of their initial directions; since the angular momenta involved in these central collisions are furthermore rather low, isotropy is to be expected.

In that case, $d\sigma_{el}/d\Omega$ is expected to look as shown in Fig. 4, and then $\sigma_{c,el}$ is 4π times the constant value of $d\sigma_{el}/d\Omega$ for large angles.

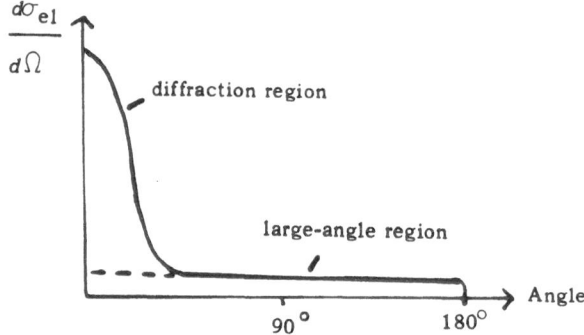

Fig. 4. Definition of $\sigma_{c,el}$ and σ_{diffr}.

We now apply the statistical model by stating

$$\sigma_{c,el} = \alpha P_0$$
$$\sigma_{c,tot} = \frac{1}{\beta}\left(\alpha \sum P_j\right) \tag{18}$$

The constants α and β are not known, and stay for the following reasons: α simply expresses that $\sigma_{c,el}$ is proportional to the probability of the elastic channel; in fact, α will drop out later. If we were to put $\sigma_{c,tot} = \alpha \sum P_j$, that is, equal to α times the sum over the probabilities of all channels, then we would underestimate $\sigma_{c,tot}$, because $\sum P_j$ is calculated from the statistical model, and consequently $\alpha \sum P_j$ gives that part of the total cross section which is due to central collisions only, whereas $\sigma_{c,tot} \approx \sigma_{inel}$ certainly contains a noncentral contribution—hence the factor β, where $\beta < 1$. We now obtain

$$\sigma_{c,\text{el}} = 4\pi \frac{d\sigma_{c,\text{el}}}{d\Omega} = \sigma_{c,\text{tot}} \beta \left(\frac{P_0}{\sum P_j} \right) \tag{19}$$

From the above considerations we should expect β to be of the order of the ratio b between central collisions and all collisions. We therefore introduce the last additional *assumption* that

$$\beta = \frac{4m^2}{E^2} \tag{20}$$

Since for large energies $\sigma_{c,\text{tot}} \longrightarrow \sigma_{\text{inel}}$, we finally obtain in the center of momentum frame

$$\frac{d\sigma_{c,\text{el}}}{d\Omega} = 2 \frac{\sigma_{\text{inel}}}{4\pi} \frac{4m^2}{E_{cm}^2} \left(\frac{P_0}{\sum P_j} \right) \qquad (pp\text{-scattering}) \tag{21}$$

where the factor 2 has been added to take into account that the two protons both contribute. If p-n scattering were considered, this factor would be absent; in π–nucleon scattering, β would have a slightly different form and the factor 2 would again be absent.

The ratio $p_0 / \sum p_j$ has been calculated numerically for various energies and the results are plotted in Fig. 5, taken from Fast and Hagedorn.[5] Expressed as a formula we obtain the result:

$$\frac{P_0}{\sum P_j} \approx \exp\left[-3.25(\sqrt{s} - 2)\right]$$

(for nucleon–nucleon scattering)

$$\frac{P_0}{\sum P_j} \approx \exp\left[-2.96(\sqrt{s} - 1.49)\right] \tag{22}$$

(for π–nucleon scattering)

Here, as well as in Fig. 5, the energy unit is the nucleon mass $m = 1$. For the nucleon–nucleon case, we obtained a straight line from the threshold up to 25 GeV primary energy; in the π–nucleon case, it is not straight in the low-energy region; formula (22) fits the straight part.

4. COMPARISON WITH RECENT EXPERIMENTS

A comparison with rather meager data for π–nucleon scattering has been made by Fast, Hagedorn, and Jones[6] and will not be considered here, since it is not conclusive.

Recent experiments with p-p large-angle (90° in the center of momentum frame) elastic scattering were done at Brookhaven between 10 and 30 GeV/c primary momentum by Cocconi *et al.*[7]

They find (this scarce information is from private communication

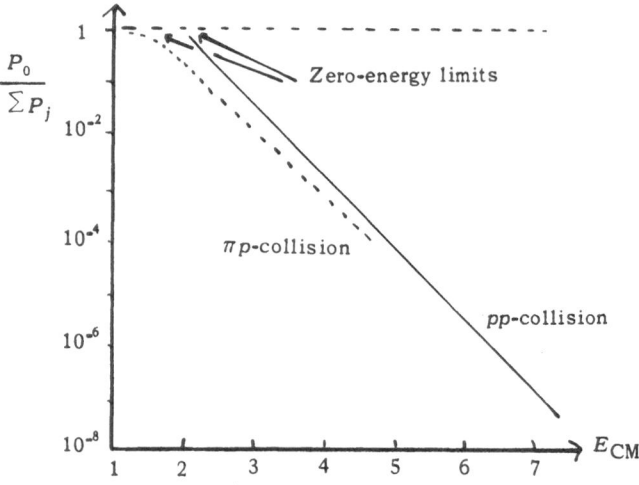

Fig. 5. The ratio $P_0 / \sum P_j$ as a function of E.

with Cocconi; the published article was not available at the time of writing this report):

> The elastic scattering is almost isotropic for $\theta > 65°$ (center of momentum) and the absolute value of $d\sigma_{\rm el}/d\Omega$ for these angles is approximately proportional to $\exp(-ap_0^{1/2})$, p_0 being the primary laboratory momentum. At $p_0 = 25$ GeV/c, they find

$$\frac{d\sigma_{\rm el}}{d\Omega} \approx 1 \times 10^{-35}\ {\rm cm}^2$$

(23)

From equation (3) it follows that for high energies $s = 2(m^2 + mE_0) \approx 2mp_0$ or $p_0^{1/2}$ is proportional to $s^{1/2}$, which agrees with equation (22).

Let us compare the value at $p_0 = 25$ GeV/c to the prediction of equations (21) and (22) (our units $\hbar = m = c = 1$).

$$p_0 = 25\ {\rm GeV}/c \qquad 26.6\ m/c = 26.6$$

$$s = 2(m^2 + m\sqrt{p_0^2 + m^2}) = 2(1 + \sqrt{(26.6)^2 + 1}) = 55.2$$

$$\sqrt{s} = E_{\rm CM} = 7.42$$

From Fig. 5 we read $p_0 / \sum p_j \approx 5 \times 10^{-8}$ and from equation (20) $\beta = 4/55.2 = 0.073$; at this energy we have furthermore $\sigma_{\rm inel} = \sigma_{\rm tot} - \sigma_{\rm el} = 40$ mb $- 10$ mb $= 3 \times 10^{-26}$ cm². With these values we obtain from (21)

$$\left(\frac{d\sigma_{c,\rm el}}{d\Omega}\right)_{\rm theory} = \frac{3 \times 10^{-26}}{2\pi} \times 0.073 \times 5 \times 10^{-8}$$

$$= 1.75 \times 10^{-35}\ {\rm cm}^2$$

which comes very near to the experimental value 1×10^{-35} cm². This result is still uncertain by a factor of the order of one, because the value of β adopted above was rather more guessed at than calculated. But even if there were disagreement by a factor larger than two, it still would not be bad, because the value of $(d\sigma_{el})/(d\Omega)$ drops from[3]

$$\left. \frac{d\sigma_{el}}{d\Omega} \right|_{\substack{\theta=0 \\ p_0 \approx 25}} \approx 100 \text{ mb} = 10^{-25} \text{ cm}^2$$

to

$$\left. \frac{d\sigma_{el}}{d\Omega} \right|_{\substack{\theta>65° \\ p_0 \approx 25}} \approx 10^{-35}$$

that is, by a factor of 10^{10}.

The recent experiments have thus shown that large-angle elastic scattering might be understood by means of the statistical model. This seems plausible, since the model should work in this case, if in any. It would be worthwhile to have further experimental information.

ACKNOWLEDGMENT

The author is grateful to G. Cocconi for assistance prior to publication of this article.

REFERENCES

1. D. Amati et al., Nuovo Cimento 26:896 (1962).
2. T. Regge, Nuovo Cimento 14:951 (1959)
3. A.N. Diddens et al., Phys. Rev. Letters 9:108 (1962); K.J. Foley et al., Phys. Rev. Letters 11:425 (1963).
4. R. Hagedorn, Nuovo Cimento 15:455 (1960); Frtschr. Phys. 9:1 (1961).
5. G. Fast and R. Hagedorn, Nuovo Cimento 27:208 (1963).
6. G. Fast, R. Hagedorn, and L.W. Jones, Nuovo Cimento 27:856 (1963).
7. G. Cocconi et al., Phys. Rev. Letters 11:499 (1963).

.

Crossing Relations and Spin States

M. Jacob[*]

MATSCIENCE
Madras, India

It is well known that in any process involving two incoming and two outgoing particles there is a symmetry between the process obtained by replacing an incoming (outgoing) particle of momentum k by an outgoing (incoming) antiparticle of momentum $-k$, *crossing symmetry*. It is the sign of the time component that decides whether a line represents an incoming particle or an outgoing antiparticle.

For a process with spinless particles, crossing symmetry (Fig. 1) implies that

$$M(p_1 q_1, p_2 q_2) = M(p_1 - q; p_2 - q) \tag{1}$$

or in terms of s, t, u, defined as usual

$$\text{Re } M(s, t, u) - \text{Re } M(u, t, s) \tag{2}$$

What happens when the particles involved have spin? The main effect is the algebraic complication introduced. For instance, consider pp-scattering, as treated by Goldberger *et al.*[1] The T-matrix may be written in terms of the Fermi invariants S, V, T, A, P and the Fierz-transformed invariants $\tilde{S}$, $\tilde{V}$, $\tilde{T}$, $\tilde{A}$, $\tilde{P}$ as

$$T = f_1(S - \tilde{S}) + f_2(T - \tilde{T}) + f_3(A - \tilde{A})$$
$$+ f_4(V - \tilde{V}) + f_5(P - \tilde{P}) \tag{3}$$

A similar set of amplitudes $\tilde{f}_i$ describes the crossed process, $p\bar{p}$-scattering. That is, if the crossed process is described by the amplitudes

---

* Visiting member of MATSCIENCE. Permanent address: Department of Physics, Saclay, France.

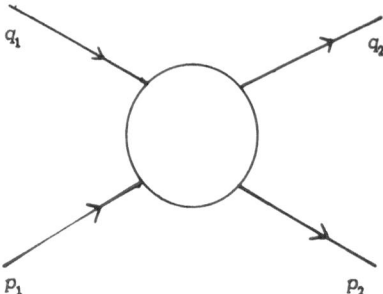

Fig. 1

$\tilde{f_i}$ defined analogously, then

$$f_i(s, t, u) = a_{ij}\tilde{f}_j(u, t, s) \tag{4}$$

where $[a_{ij}]$ is a numerical matrix.[1]

The helicity amplitudes φ_i, $i = 1, \ldots, 5$, may be defined in the usual way (see Table I). They are related to the f_i by $\varphi = \Gamma f$. In a similar way $\bar\varphi = \bar\Gamma \bar f$, where Γ, $\bar\Gamma$ are known matrices which are functions of the energy and momenta. The $\bar\varphi_i$ are the helicity amplitudes for the crossed process.

Conservation of parity and time reversal invariance are used to reduce the number of independent pp-amplitudes to five. In the crossed $p\bar p$-channel, we use parity conservation and charge conjugation invariance.

Thus, the crossing relations for the helicity amplitudes read

$$\varphi = [\Gamma a \bar\Gamma^{-1}]\bar\varphi \tag{5}$$

which cannot be used as such to continue from one channel to the other, because this matrix has branch-point singularities. However, the corresponding relation obtained with a new set of amplitudes

$$\psi_i = \frac{W}{2}\varphi_i \qquad i = 1, 2, 3, 4 \qquad W = \sqrt{s} \tag{6}$$

$$\psi_5 = \frac{2m}{\sin\theta}\varphi_5$$

does not involve branch-point singularities:

$$\psi_i(s, t, u) = C_{ij}(s, t, u,)\bar\psi_j(u, t, s) \tag{7}$$

We then go from the ψ to the φ in each channel.

We therefore obtain a direct relation between helicity amplitudes, in both channels, through elimination of the invariant amplitudes.

In order to illustrate that this elimination is not a compulsory step, we shall consider the simpler case of π^- nucleon scattering.

From here on, we shall follow some recent work of Trueman and Wick[2] to obtain the crossing relations directly for the helicity amplitudes.

In this case, the helicity amplitudes for the direct process $G_{\lambda_f \lambda_i}$ and for the crossed process $F_{\lambda \bar{\lambda}}$ may be related to the invariant amplitudes A, B by

$$G_{++} = \cos \frac{\theta}{2} \left(A + \frac{k^2 + \epsilon\omega}{m} B \right) \tag{8a}$$

$$G_{+-} = \sin \frac{\theta}{2} \left(\frac{\epsilon}{m} A + \omega B \right) \tag{8b}$$

where ω, ϵ are the π and N center of mass energies in the direct channel, and

$$F_{++} = -\frac{p}{m} A + q \cos \bar{\theta} \cdot B \tag{9a}$$

$$F_{+-} = \frac{Eq}{m} \cos \bar{\theta} \cdot B \tag{9b}$$

where $\cos \bar{\theta} = (s - u)/(u\, pq)$. (All variables here refer to the crossed process.) [We have written $T = -A + (i\gamma \cdot Q)B$.] $G_{\lambda\mu}$ may be obtained in terms of $F_{\lambda\bar{\lambda}}$, provided we specify the path followed in continuing $\cos \bar{\theta}, p, q$, etc., from the physical region of the direct process to that of the crossed process. We find

$$G_{++} = \frac{2}{\Lambda^{1/2}} (mq \sin \bar{\theta} F_{++} + E(p - q \cos \bar{\theta})F_{+-})$$

$$G_{+-} = \frac{2}{\Lambda^{1/2}} (E(p - q \cos \bar{\theta})F_{++} - mq \sin \bar{\theta} F_{+-}) \tag{10}$$

with

$$\Lambda = \Lambda(s^{1/2}, m, \mu) \tag{11}$$

Table I. Helicity Amplitudes

f \ i	$++$	$+-$	$-+$	$--$
$++$	ϕ_1	ϕ_5	$-\phi_5$	ϕ_2
$+-$	$-\phi_5$	ϕ_3	ϕ_4	$-\phi_5$
$-+$	ϕ_5	ϕ_4	ϕ_3	ϕ_5
$--$	ϕ_2	ϕ_5	$-\phi_5$	ϕ_1

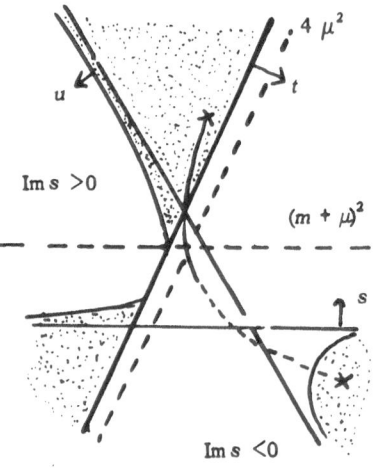

Fig. 2

and

$$\Lambda(xyz) = (x + y + z)(x + y - z)(x - y + z)(x - y - z) \qquad (12)$$

On using the path shown in Fig. 2, we obtain the simple orthogonal relation

$$\begin{bmatrix} G_{++} \\ G_{+-} \end{bmatrix} = \begin{bmatrix} \sin x & \cos x \\ \cos x & -\sin x \end{bmatrix} \begin{bmatrix} F_{++} \\ F_{+-} \end{bmatrix} \qquad (13)$$

where

$$\tan x = \frac{mq \sin \bar{\theta}}{E(p - q \cos \bar{\theta})}$$

We have obtained the relation between the helicity amplitudes $G_{\lambda\mu}$ and $F_{\lambda\bar{\lambda}}$ by first relating each of these to the scalar amplitudes.

We shall now show how the crossing relations between the helicity amplitudes may be obtained *without* first relating them to the invariant amplitudes.

So far, we have defined the helicity amplitudes in the center of mass system of each process. However, these amplitudes may be defined in any reference frame.

We shall denote states with a definite momentum p and helicity λ by

$$|P, \lambda\rangle \qquad (14)$$

We have

$$G_{\mu,\lambda} = \langle P_2, \mu | T | P_1, \lambda \rangle \qquad (15)$$

Since T is Lorentz-invariant,

$$T = L^{-1}TL \qquad (16)$$

therefore

$$G_{\mu,\lambda} = \langle P_2, \mu \mid L^{-1}TL \mid P_1, \lambda \rangle$$
$$= \langle L\{P_2, \mu\} \mid T \mid L\{P_1, \lambda\} \rangle \tag{17}$$

We must now ask what the effect of a Lorentz transform L is on a state $\mid P, \lambda \rangle$.

If L is a pure velocity transformation along some direction, then it does not commute with the helicity operator unless the Lorentz transformation is in the same direction as the momentum P in the state $\mid P, \lambda \rangle$.

In general, we have

$$L \mid P_1, \lambda \rangle = \sum_{\lambda'} U_{\lambda'\lambda}(l, p_1) \mid p_1, \lambda' \rangle \tag{18}$$

with $P_1 \longrightarrow p_1$ through the transformation L; this is written

$$p_1 = L P_1$$

and

$$\langle P_2, \mu \mid L^{-1} \equiv \langle P_2, \mu \mid L^+ = \sum_{\mu'} U^*_{\mu'\mu}(l, P_2) \langle p_2, \mu' \mid \tag{19}$$

Now denote by $\mid p_1, \lambda' \rangle$, $\mid p_2, \mu' \rangle$ states with momentum p_1, p_2 and helicity λ', μ', respectively, in the center of mass frame, and by $\mid P_1, \lambda \rangle$, $\mid P_2, \lambda \rangle$ states in the frame obtained by applying the Lorentz transform L to the center of mass frame. We have

$$U^*_{\mu'\mu}(l, p_2) = U^{-1}_{\mu\mu'}(l, p_2) = U_{\mu\mu'}(p_2, l^{-1}) \tag{20}$$

So we now have the relation between the helicity amplitudes in the center of mass frame and any frame:

$$\langle P_2, \mu \mid T \mid P_1, \lambda \rangle = \sum_{\lambda'\mu'} U_{\mu\mu'}(l^{-1}, P_2) U_{\lambda'\lambda}(P_1, l) \langle p_2, \mu' \mid T \mid p_1, \lambda' \rangle \tag{21}$$

The $U_{\mu'\mu}$ are rotation matrix elements. On applying a Lorentz transformation, not only does the momentum change, but there also is a rotation of the spin.

We may see this by making the following sequence of transformations:

Start with the states at rest in the center of mass frame (with 4-momentum p_1^0) and make a Lorentz transformation to obtain the state with momentum P_1:

$$P_1 = h(P_1) p_1^0 \tag{22a}$$

Now make the Lorentz transformation l to give a state with 4-momentum p_1.

$$p_1 = l P_1 = l h(p_1) p_1^0$$

Finally, make the Lorentz transform on the state p_1 that brings it back to rest, i.e., $h^{-1}(p_1)$ or $h^{-1}(l\,P_1)$

$$p_1^0 = h^{-1}(lP_1)lh(P_1)p_1^0$$

The transformation $h^{-1}(l\,P_1)\,l\,h\,(P_1)$ must then be a rotation.

To get the crossing relation between the center of mass amplitudes for the direct and crossed processes, we must continue the scattering amplitude and the expressions for the rotation matrix elements $U(l\,p)$ from positive time-like values of p_1 to negative time-like values of p_1, so as to obtain

$$G_{\lambda'\mu'}(p_2, -q_2; -p_1, q_1) \qquad (23)$$

which are, up to a phase factor, just the helicity amplitudes for the crossed process $F_{\lambda'\mu'}$.

When the particles involved have spin, both the momentum and the spin component are reversed in crossing. Therefore, the helicity is left unaltered.

The relation obtained for pion-nucleon scattering is then found to be

$$G_{\mu\lambda}(s, t) = \sum_{\mu'\lambda'} d_{\mu'\mu}^{1/2}(\pi - x)d_{\lambda'\lambda}^{1/2}(x)(-)^{\mu'-\lambda'}F_{\mu'\lambda'}(s, t) \qquad (24)$$

Using the well-known expression of the d-functions, we get equation (10). The significance of the angle x has, however, been clarified, and the transformation matrix obtained directly. In order to do so we have to calculate the rotation angle associated to each Lorentz transformation. For instance, we write

$$U_{\lambda'\lambda}(P, l) = d_{\lambda'\lambda}^{1/2}(x)$$

This simple form will hold as long as all the Lorentz transformations are performed in one plane, which can always be done.

The problem of finding x can be transformed into a problem of spherical trigonometry in the following way:

It is usual to associate to a special Lorentz transformation, along the z-axis, say, a rotation by an angle $i\beta$ in the (z, it) plane. β is defined by $\tanh \beta = v/c$. In other words, the Lorentz transformation is represented by an arc β on a unit circle of radius i.

If we now combine two successive special Lorentz transformations, we obtain the resultant by adding the two pertinent arcs. If these two Lorentz transformations are performed along different axes, we add the two arcs defined on a sphere of radius i. The angle between the two arcs is, of course, real and equal to the angle between the two velocity vectors.

For the sake of argument, let us draw a sphere and assume that we go from an initial velocity represented by a point A to a different velocity, represented by a point B, through a special Lorentz transformation corresponding to the angle $i\beta_1$. In a similar way, we go from B to C by a second transformation defined by the angle $i\beta_2$.

The Lorentz transformation leading from C to A is defined by an angle $i\beta$ and, using the hyperbolic function instead of the trigonometric function of an imaginary angle, we have

$$\cosh \beta = \cosh \beta_1 \cosh \beta_2 - \sinh \beta_1 \sinh \beta_2 \cos \theta \qquad (25)$$

Angle θ is the angle between the velocity vectors of the special transformations $A \rightarrow B$ and $B \rightarrow C$. There are two other similar relations. It then also follows that

$$\frac{\sin \theta}{\sinh \beta} = \frac{\sin \varphi_1}{\sinh \beta_1} = \frac{\sin \varphi_2}{\sinh \beta_2} \qquad (26)$$

We now assume that B represents the rest system of particle 1, A and C the motion of the center of mass system for the reactions.

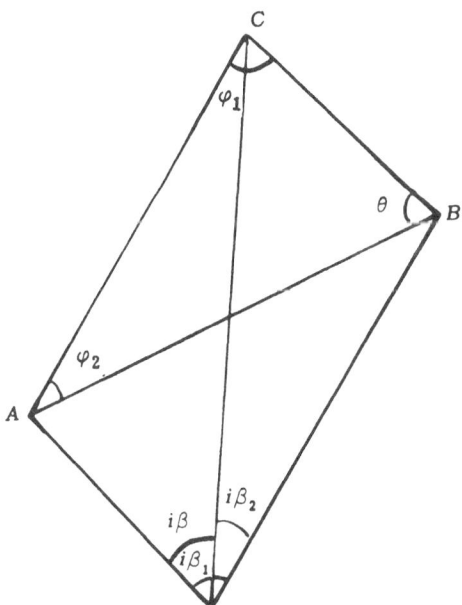

Fig. 3

θ will then be the rotation angle when going from one system to the other. It is the angle of the velocity vectors of the two systems as seen from the rest system of the particle.

Using equations (24) and (25), we obtain x_1 with

$$\sin x_1 = \frac{2m \sin \bar{\theta}}{\Lambda^{1/2}} \quad \text{and} \quad \cos x_1 = \frac{2E(p - q \cos \bar{\theta})}{\Lambda^{1/2}}$$

In a similar way, one obtains for particle 2 and angle $x_2 = \pi - x_1$ the representation given in Fig. 3. The phase factor previously mentioned can be shown to be $(-)^{\lambda' - \mu'}$.

REFERENCES

1. Goldberger et al., *Phys. Rev.* **120**: 2250 (1960).
2. Trueman and Wick, *Ann. Phys.* **26**: 322 (1964).

The Multiperipheral Model for High-Energy Processes

K. VENKATESAN

MATSCIENCE
Madras, India

The multiperipheral model developed by Amati, Fubini, and others to treat high-energy scattering and production processes is a generalization to very high energies of the idea of peripheral interactions, which seem to be particularly dominant in the 1 to 3 BeV region. Here, we shall be concerned only with the behavior of high-energy cross sections for two-particle processes.

The basis of the peripheral model is the strip approximation to the Mandelstam representation. Let us consider for simplicity a reaction involving only a single invariant amplitude $F(s, t, u)$ which is a function of the usual Mandelstam variables s, t, and u. (An instance of such an interaction is the photoproduction of pions on pions, which involves only a single isotopic spin amplitude.) The double dispersion relation for $F(s, t, u)$ can be written as

$$F(s, t, u) = \text{pole terms} + \frac{1}{\pi^2} \iint ds' dt' \frac{F_{13}(s', t')}{(s' - s)(t' - t)}$$
$$+ \frac{1}{\pi^2} \iint ds' du' \frac{F_{12}(s', u')}{(s' - s)(u' - u)} \tag{1}$$
$$+ \frac{1}{\pi^2} \iint dt' du' \frac{F_{23}(t', u')}{(t' - t)(u' - u)}$$

where F_{13}, F_{12}, and F_{23} are the double spectral functions. Fixing the momentum transfer variable t, this can be written as

$$F(s, t, u) = \text{poles} + \frac{1}{\pi} \int ds' \frac{A_1(s', t)}{s' - s}$$
$$+ \frac{1}{\pi} \int du' \frac{A_2(u', t)}{u' - u} \tag{2}$$

where the absorptive parts A_1 and A_2 are for the s- and u-channels, respectively. A_1 obeys the dispersion relation

$$A_1(s, t) = \int \frac{F_{13}(s, t')}{t' - t} dt' + \frac{F_{12}(s, u')}{u' - u} du' \tag{3}$$

Since we shall be interested in high values of the energy variable s, the fact that the square of the momentum transfer t has small values at these energies would imply, if we use the relation $s + t + u$ equals the sum of the squares of masses of the initial and final particles, that u is large and negative, so that the denominators in the second terms of equations (2) and (3) will be large and positive. Therefore, we omit these terms from further consideration. (If, however, large momentum transfers are also involved at high energies, as recent cosmic ray experiments seem to indicate, neglect of the u-dependent terms would not be justified.)

To obtain an expression for the double spectral function, we start from the unitarity condition

$$A_1(s, z_1) = \int d^2 q F^*(s, z_3) F(s, z_2) \tag{4}$$

for a scattering process. $z_1 = \cos \theta$ is the angle between the directions of the initial and final set of particles in the center of mass. z_2 is the angle between the directions of the initial set of particles and those in the intermediate state (which here is the lowest mass two-particle state). z_3 is the angle between the directions of the intermediate and final set of particles. These angles can also be re-expressed in terms of the momentum transfer variables t, t_2, and t_3.

Using the fixed energy-dispersion relations for the F and performing the integration over the intermediate angles, equation (4) can be written as

$$A_1(s, z_1) = \iint \frac{dz_2' dz_3'}{k^{1/2}} \log \frac{z_1 - z_2' z_3' + k^{1/2}}{z_1 - z_2' z_3' - k^{1/2}}$$
$$\times A_3^*(s, z_3') A_3(s, z_2') \tag{5}$$

where

$$k = z_1^2 + z_2'^2 + z_3'^2 - 2z_1 z_2' z_3' - 1 \tag{6}$$

The function $(1/k^{1/2}) \log (z_1 - z_2' z_3' + k^{1/2})/(z_1 - z_2' z_3' - k^{1/2})$ has two branch cuts in the z_1 (or equivalently the t) plane, one on the right-hand and the other on the left-hand side. Identifying the right-hand cut with the first term on the right-hand side of equation (3), we find that

$$F_{13}(s, z_1) = \iint dz_2 dz_3 K(s, z_1, z_2, z_3) A_3^*(s, z_3) A_3(s, z_2) \tag{7}$$

where

$$K(s, z_1, z_2, z_3) = -\frac{1}{k^{1/2}}$$

for $z_1 > z_3 z_3 - (z_2^2 - 1)^{1/2}(z_3^2 - 1)^{1/2}$, and

$$K(s, z_1, z_2, z_3) = 0$$

(8)

for $z_1 < z_2 z_3 - (z_2^2 - 1)^{1/2}(z_3^2 - 1)^{1/2}$

In terms of the t variables and for all s, equation (8) could be written as

$$K(s, t, t_2, t_3) = 0$$

unless

$$t^{1/2} > t_2^{1/2} + t_3^{1/2} \qquad (9)$$

For any particular s, the restrictions on the variables can be strengthened.

The function F_{13} [equation (7)] was obtained by using the elastic unitarity condition in the s-channel (that is, we have retained only the lowest-mass two-particle state in the intermediate state). If we write the equation for the absorptive part as

$$A_1(s, t) = \int \frac{F_{13}^{el}(s, t')}{t' - t} \, dt' + \int \frac{F_{13}^{inel}(s, t')}{t' - t} \, dt' \qquad (10)$$

then the function $F_{13}(s, t)$ in equation (7) represents precisely the function $F_{13}^{el}(s, t)$. To obtain $F_{13}^{inel}(s, t)$, we make use of the strip approximation of Chew and Frautschi, which states that

$$F_{13}^{inel}(s, t) = F_{13}^{el}(t, s) \qquad (11)$$

The approximation is based on the idea that the strips of the double spectral functions lying between the thresholds for the two-particle and next-lowest-mass three-particle states in each of the three channels s, t, and u, which happen to lie contiguous to the physical region, give the major contributions to the processes. Thus, $F_{13}^{inel}(s, t)$ is determined by evaluating the double spectral function using the elastic unitarity condition in the t-channel. Since for most reactions the low-energy region is dominated by a few resonances, we could write the absorptive part in the s-channel as

$$A_1(s, t) = A^R(s, t) + \int\int dx \, dy \, K(t, s, x, y) A_1^*(t, x) A_1(t, y) \qquad (12)$$

The second term representing the inelastic contribution is given in Fig. 1.

The squares of the total center of mass energies of the blobs x and

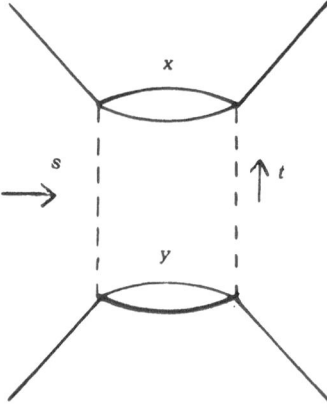

Fig. 1

y are such that they obey the condition (9), that is, $s^{1/2} > x^{1/2} + y^{1/2}$. For high energies, where this condition may not obtain, we divide the blobs by increasing the number of rungs in the t direction (as shown in Fig. 2) such that the "masses," x', x'', ..., y' obey the above restriction pair-wise. This is the multiperipheral chain, which, considered as a Feynman diagram, can be connected with a chain having one less rung by the expression

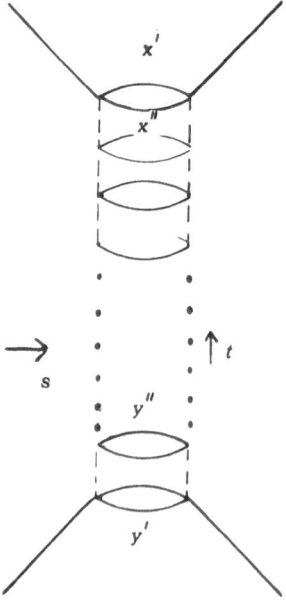

Fig. 2

$$A_n(s, u_1, u_2, t) = \int ds_0 A^R(s_0, t)$$

$$\times \int\int\int ds'\, du_1'\, du_2'\, K(s, s', t, s_0, u_1, u_2, u_1', u_2') \qquad (13)$$

$$\times \frac{A_{n-1}(s', t, u_1', u_2')}{(u_1' + \mu^2)(u_2' + \mu^2)}$$

where we have divided the rungs into two parts (as shown in Fig. 3),

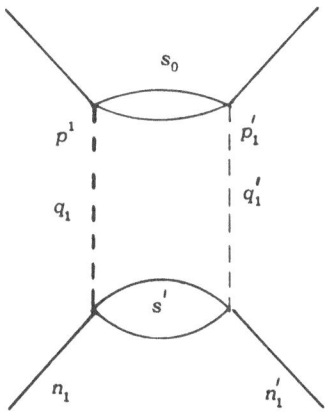

Fig. 3

one containing a single rung with the exchanged square of the "mass" s_0 represented by the resonant amplitude $A^R(s_0, t)$, and the other containing $(n-1)$ rungs with the exchange square of the "mass" s'. The values u_1 and u_2 are the off-mass shell values of the squares of the four-momenta q_1 and q_1' of the intermediate pions in the t-channel:

$$K(s, s', t, s_0, u_1, u_2, u_1', u_2') = \int d^4 Q\, \delta[(P - Q)^2 - s_0]\delta[(Q + \Delta)^2 + u_1']$$
$$\times \delta[(Q - \Delta)^2 + u_2']\delta[(Q + N)^2 - s']$$
$$= \frac{\theta(J)}{J^{1/2}} \qquad (14)$$

where the determinant J is given by

$$J = \begin{vmatrix} -\frac{t}{2} & \frac{u_1'-u_2'}{2} & \frac{u_1-u_2}{2} & 0 \\[2mm] \frac{u_1'-u_2'}{2} & u_1'+u_2'+\frac{t}{2} & s_0+\frac{u_1+u_2+u_1'+u_2'+t}{2} & s'-\mu^2+\frac{u_1'+u_2'+t}{2} \\[2mm] \frac{u_1-u_2}{2} & \frac{u_1+u_2+u_1'+u_2'+t}{2}+s_0 & u_1+u_2+\frac{t}{2} & s-\mu^2+\frac{u_1+u_2+t}{2} \\[2mm] 0 & s'-\mu^2+\frac{u_1'+u_2'+t}{2} & s-\mu^2+\frac{u_1+u_2+t}{2} & 2\mu^2-\frac{t}{2} \end{vmatrix}$$

$$(15)$$

and

$$Q = \frac{q_1 + q_1'}{2} \qquad P = \frac{p_1 + p_1'}{2}$$

$$N = \frac{n_1 + n_1'}{2} \qquad \Delta = \frac{q_1 - q_1'}{2} = \frac{p_1 - p_1'}{2} = \frac{n_1' - n_1}{2} \qquad (16)$$

$$u_1 = -(P + \Delta)^2 \qquad u_2 = -(P - \Delta)^2$$

$$s = (p_1 + n_1)^2 \qquad s_0 = (P - Q)^2 \qquad s' = (N + Q)^2$$

At high energies, the resonant contribution to the process is negligible, so that the first term on the right-hand side of equation (12) can be neglected. Also, when s is large compared to u_1, u_2, u_1', u_2', and s_0, we have

$$\frac{\theta(J)}{J^{1/2}} \sim \frac{\theta[H(\xi, \xi_1, \xi_2)]}{s[H(\xi, \xi_1, \xi_2)]^{1/2}} = \frac{2}{s} T(\xi, \xi_1, \xi_2). \qquad (17)$$

where

$$\xi = -t\left(1 - \frac{s'}{s}\right)$$

$$\xi_1 = u_1' - u_1 \frac{s'}{s} - \frac{s_0 s'}{s - s'}$$

$$\xi_2 = u_2' - u_2 \frac{s'}{s} - \frac{s_0 s'}{s - s'}$$

$$H(\xi, \xi_1, \xi_2) = \tfrac{1}{4}(-\xi^2 - \xi_1^2 - \xi_2^2 + 2\xi\xi_1 + 2\xi\xi_2 + 2\xi_1\xi_2) \qquad (18)$$

Summing over all the multiperipheral graphs (that is, over n), we obtain an integral equation for the absorptive part. (We omit the suffix 1 on the absorptive part from now on.)

$$A(s, t, u_1, u_2) = \int ds_0\, A^R(s_0, t)$$

$$\times \iiint ds'\, du_1'\, du_2' \qquad (19)$$

$$\times \frac{2T(\xi, \xi_1, \xi_2)\, A(s', t, u_1', u_2')}{s(u_1' + \mu^2)(u_2' + \mu^2)}$$

We observe that equation (19) is invariant under the scale transformation

$$s \longrightarrow cs \qquad s' \longrightarrow cs'$$

so that we can try a solution of the form

$$A(s, t, u_1, u_2) = s^{\alpha(t)} \phi(u_1, u_2, t) \qquad (20)$$

where $\phi(u_1, u_2, t)$ satisfies the integral equation

$$\phi(u_1, u_2, t) = \int ds_0 \, A^R(s_0, t)$$

$$\times \int_0^1 dx \, x^{\alpha(t)}$$

$$\times \int\int \frac{2du_1' \, du_2'}{(u_1' + \mu^2)(u_2' + \mu^2)} \, T(\xi, \xi_1, \xi_2) \phi(u_1', u_2', t)$$

with

$$x = \frac{s'}{s} \qquad (21)$$

Expression (20) shows the characteristic Regge behavior of the absorptive amplitude at high energies. It has followed as a consequence of the scale invariance of the integral equation for the absorptive part (and not through a process of analytic continuation).

Equation (21) is a homogeneous Fredholm integral equation which defines an eigenvalue problem. There are solutions for specific values of $\alpha(t)$ if we treat t as a parameter. To obtain limits on the values $\alpha(t)$ can assume and to obtain an equation for the Regge trajectory, we start from equation (19) and notice that large contributions to the integrals over u_1' and u_2' arise for small values of u_1' and u_2' which are very near $-\mu^2$. Hence we can set

$$A(s', t, u_1', u_2') \simeq A(s', t, -\mu^2, -\mu^2)$$
$$\simeq s'^{\alpha(t)} \phi(t)$$

so that

$$1 = \int ds_0 \, A^R(s_0, t) \int_0^1 dx \int\int du_1' \, du_2'$$
$$\times \frac{x^{\alpha(t)} T(\xi, \xi_1, \xi_2)}{(u_1' + \mu^2)(u_2' + \mu^2)} \qquad (22)$$

Putting $A^R(s_0, t) = g^2 f(s_0)$ and using the identity

$$T(\xi, \xi_1, \xi_2) = 2 \int d^2q \, \delta[(p + q)^2 - \xi_1] \delta[(p - q)^2 - \xi_2]$$

where p is a two-dimensional vector with

$$4p^2 = -t(1 - x)$$

we obtain on performing the u_1', u_2' and the angle integration (over d^2q), using Feynman parameterization, the relation

$$1 = g^2 \int ds_0 f(s_0) \int_{-1}^1 dz \int_0^1 dx \, x^{\alpha}$$
$$\times \left[-\frac{t}{4}(1 - z^2)(1 - x) + \frac{s_0 x}{1 - x} + \mu^2(1 - x) \right]^{-1} \qquad (23)$$

which gives an equation for the trajectory of $\alpha(t)$. By differentiating
equation (23), it can be shown that $d\alpha/dt > 0$ for $t \leq 4$ and that α (t)
becomes complex if $t \geq 4$. This can also be seen from the following:
Write equation (23) in the form

$$1 = g^2 \int ds_0 f(s_0) \int_{-1}^{1} \frac{dz}{-\frac{1}{4}t(1-z^2)+\mu^2} \int_0^1 dx \frac{x^\alpha - x^{\alpha+1}}{x^2 - 2x\cos\lambda + 1} \tag{24}$$

where

$$\cos\lambda = 1 - \frac{s_0}{2[-\frac{1}{4}t(1-z^2)+\mu^2]}$$

Integrating over x, we have

$$1 = g^2 \int ds_0 f(s_0) \int_{-1}^{1} \frac{dz}{-\frac{1}{4}t(1-z^2)+\mu^2}$$
$$\times \sum_{n=1}^{\infty} \frac{\sin n\lambda}{\sin\lambda}\left(\frac{1}{n+\alpha} - \frac{1}{n+\alpha+1}\right) \tag{25}$$

There are singularities whenever $\alpha(t)$ takes negative integral values.
The series converges for $\alpha(t) > -1$. The denominator $-\frac{1}{4}t(1-z^2)$
$+ \mu^2$ begins to vanish when $t = 4\mu^2$, and a term $+i\epsilon$ has to be added
if the equation is to have meaning. This means that for $t > 4\mu^2$,
$\alpha(t)$ will become complex.

We can determine the real part $D(s, t)$ of the amplitude from the
imaginary part $A(s, t)$ by means of a fixed momentum-transfer dispersion relation. If the asymptotic form of $A(s, t)$ is as given by equation
(20) we must write for $D(s, t)$ a dispersion relation with m subtractions,
where m is the least integer greater than $\alpha(t)$. Therefore

$$D(s, t) = \frac{s^m}{\pi} P \int \frac{A(s', t)ds'}{s'^m(s'-s)} + \frac{u^m}{\pi} P \int \frac{A(u', t)du}{u'^m(u'-u)}$$
$$+ \text{ polynomials in } s \text{ and } u \tag{26}$$
$$(\text{of maximum power } [m-1])$$

In the asymptotic region, the polynomial terms can be neglected.
We have included the crossed-channel contribution also, since neglect
of it will lead to a contradiction, as shown below. The lower limit of the
integrals can be taken as 0 (since there is no infrared divergence).

Since in the asymptotic limit $u = -s$, the asymptotic form of
$D(s, t)$ can be written as

$$D(s, t) = \frac{s^m\phi(t)}{\pi} P \int_0^\infty \frac{ds'}{s'^{m-\alpha}(s'-s)}$$
$$+ (-1)^m s^m \frac{\phi(t)}{\pi} P \int_0^\infty \frac{ds'}{s'^{m-\alpha}(s'+s)} \tag{27}$$

The crossed term is equal to

$$D_{\text{cross}}(s, t) = (-1)^m \frac{\phi(t)}{\pi} s^{\alpha(t)} \int_0^\infty dx \frac{x^{\alpha-m}}{1+x}$$

$$= (-)^m \frac{\phi(t)s^{\alpha(t)}}{\sin \pi(\alpha + 1 - m)} = \frac{-\phi(t)s^{\alpha(t)}}{\sin \pi\alpha(t)} \tag{28}$$

$$D_{\text{uncrossed}}(s, t) = -\phi(t)s^{\alpha(t)} \cot \pi\alpha(t) \tag{29}$$

Therefore

$$D(s, t) = -\phi(t)s^{\alpha(t)} \cot \frac{\pi\alpha(t)}{2} \tag{30}$$

Thus, $D(s, t)$ has poles for any even integral value of $\alpha(t)$ and vanishes when $\alpha(t)$ is an odd integral. This is just the statement of the Pauli principle for the scattering of identical particles in the t-channel. If we had not taken the crossed channel into account, we would have obtained a real part proportional to $\cot \pi\alpha(t)$ with poles for every integral value of $\alpha(t)$.

The total amplitude $F(s, t)$ thus has the asymptotic form

$$F(s, t) = -\phi(t)s^{\alpha(t)} \frac{\exp[\{-i\pi\alpha(t)\}/2]}{\sin[\pi\alpha(t)/2]} \tag{31}$$

If we take the case when the scattering amplitudes are antisymmetric under the exchange of s and u, we have

$$D(s, t) = \phi(t)s^{\alpha(t)} \tan\left[\frac{\pi\alpha(t)}{2}\right]$$

$$F(s, t) = i\phi(t)s^{\alpha(t)} \frac{\exp[-i\pi\alpha(t)/2]}{\cos[\pi\alpha(t)/2]} \tag{32}$$

Thus the amplitude has poles for odd integral values of $\alpha(t)$.

Equations (31) and (32) are identical to those conjectured by Frautschi *et al.*, under the assumption that Regge's results for potential scattering may be extended to the relativistic case.

Using the optical theorem, we have for the asymptotic form of the total cross section

$$\sigma_{\text{in}}(s) = \frac{A(s, 0)}{s} = \phi(0)s^{\alpha(0)-1} \tag{33}$$

which will have a constant behavior if $\alpha(0) = 1$. The elastic contribution which has to be added to equation (33) is given by

$$\sigma_{\text{el}}(s) = \frac{1}{32\pi} \int_{-1}^{1} \frac{|F|^2}{s} d\cos\theta$$

$$= \frac{1}{16\pi} \int_{-t_{\text{max}}}^{0} s^{2(\alpha(t)-1)} \frac{\phi^2(t)dt}{\sin^2[\pi\alpha(t)/2]} \tag{34}$$

$$\propto \frac{s^2[\alpha(0)-1]}{\log s}$$

where use has been made of the fact that $\alpha(t)$ is an increasing function so that $\alpha(0)$ is the maximum value of $\alpha(t)$ in the range of integration. $t_{\max}$ is the maximum value of the momentum transfer which it takes when $\cos \theta = -1$.

$$
\begin{aligned}
\frac{d\sigma_{\text{el}}(s, t)}{dt} &= \frac{|F(s, t)|^2}{16\pi s} \\
&= \frac{\phi^2(t)}{16\pi} s^{2(\alpha(t)-1)} \left\{ \begin{array}{l} \sin^{-2}\left[\dfrac{\pi\alpha(t)}{2}\right] \\[2mm] \cos^{-2}\left[\dfrac{\pi\alpha(t)}{2}\right] \end{array} \right\} \\
&= \frac{\phi^2(t)}{16\pi} \exp\left[2\alpha'(0) \log s\right] \left\{ \begin{array}{l} \sin^{-2}\left[\dfrac{\pi\alpha(t)}{2}\right] \\[2mm] \cos^{-2}\left[\dfrac{\pi\alpha(t)}{2}\right] \end{array} \right\}
\end{aligned} \tag{35}
$$

showing the characteristic shrinking of the diffraction pattern with energy increase [as $\alpha'(0)$ is negative].

We mentioned that $\alpha(t) > -1$ to get a meaningful solution for the integral equation, and in order for the Pomeranchuk theorem to be satisfied $\alpha(0) = 1$. Thus for some (negative) value of t in the physical region $\alpha(t) = 0$, giving rise to an unphysical particle of imaginary mass which can travel with a velocity greater than that of light. In the present model this unwanted "ghost" cannot be eliminated by saying that its coupling constant to physical particles is zero since $\phi(t)$ is positive for all negative values of t. An alternative way of eliminating the ghost and at the same obtaining an upper limit for the values of $\alpha(t)$ (the Froissart limit is $\alpha(t) \leq 1$ for $t \leq 0$) would be to consider the unitarity in the s-channel. What has been done until now is to consider that in the unitarity condition

$$
A(s, t) = \tfrac{1}{2}(F^{\text{inel}*} F^{\text{inel}} + F^{\text{el}*} F^{\text{el}}) \tag{36}
$$

only the inelastic part is dominant, which leads to the Regge behavior (equation 20) for the absorptive part $A(s, t)$. We can use this as the next approximation by substituting equation (31) or (32) into the second term of equation (36).

Thus, we obtain as the first correction at nonzero angles the expression

$$
\begin{aligned}
A(s, t) &= \frac{2}{(8\pi)^2 s} \int_{-\infty}^{0} dt' \int_{-\infty}^{0} dt'' \, F^{\text{inel}*}(s, t') \\
&\quad \times F^{\text{inel}}(s, t'') T(t, t', t'') \\
&= \int d\xi \, s^\xi \rho(t\xi)
\end{aligned} \tag{37}
$$

where

$$\rho(t\xi) = \int_{-\infty}^{0} dt' \int_{-\infty}^{0} dt'' \delta[\xi + 1 - \alpha(t') - \alpha(t'')]$$
$$\times T(t, t', t'')\phi(t')\phi(t'') \frac{\cos\{(\pi/2)[\alpha(t') - \alpha(t'')]\}}{\sin[(\pi/2)\alpha(t')]\sin[(\pi/2)\alpha(t'')]} \tag{38}$$

From the fixed t-dispersion relation, we have

$$F_1(s, t) = \int_{\xi_m(t)}^{\xi_M(t)} s^\xi \rho(t\xi) \left[-\cot\frac{\pi\xi}{2} + 1 \right] d\xi \tag{39}$$

We see that while $F_0 = F^{\text{inel}}$ showed an asymptotic behavior of $s^{\alpha(t)}$, $F_1(s, t)$, the next correction, shows the behavior of a superposition of powers of s up to a maximum of $\xi_M(t)$—that is, while F_0 showed a Regge-pole behavior, F_1 displays a Regge-cut behavior.*

It can be shown that

$$\xi_M(0) \geq \xi_M(t) > \alpha(t) - \alpha(0) - 1 \qquad \text{for } t < 0$$

while

$$\xi_M(0) = 2\alpha(0) - 1 \tag{40}$$

Values of $\alpha(0)$ greater than 1 would thus be inconsistent with the present model. Any further iteration imposed by unitarity in the s-channel would increase the power of s. This would mean that the high-energy amplitude would increase faster than any power law. Thus, the limitation on the value of $\alpha(t)$ is the Froissart limit mentioned earlier.

For $\alpha(0) = 1$, $\xi_M(t) \geq \alpha(t)$ where the equality sign holds for $t = 0$. This means that for $t < 0$ the maximum power of s appearing in F_1 is always higher than $\alpha(t)$ except for $t = 0$, when both coincide. Thus, except for the forward direction, the asymptotic behavior of F_1 (the cut) dominates F_0 (the Regge pole). For $\alpha(0) < 1$, the relative importance of F_0 and F_1 varies. The crossed term involving these two contributions can lead to a solution $s^{\beta(t)}$ with, however, a discrete exponent $\beta(t)$ different from $\alpha(t)$. This is a typical "renormalization" effect of the trajectory of the pole $\alpha(t)$, which, it is expected, might somehow eliminate the ghost problem.

It has been experimentally observed that whereas there is some shrinking in the diffraction peak for proton–proton scattering and K^+-nucleon scattering, there is no such shrinking for the pion–nucleon case. To see whether this dependence on the energy and the process of

---

* Mandelstam has shown that these particular cuts are cancelled by other contributions to the scattering amplitudes.[9]

the shrinking of the diffraction peak can be explained by the present theory, let us try to solve the equation for the absorptive part, with the Pomeranchuk Regge pole as the inhomogeneous term. Writing

$$\phi(t)s^{\alpha(t)} \approx as\left(\frac{s}{s_0}\right)^{\alpha(0)t} = as\,e^{b(s)t} \tag{41}$$

where $b(s) = [\alpha(0) \log s]/s_0$, a is the asymptotic value of the inelastic cross section, and s_0 a constant, we have as the equation to be solved

$$A(s, t) = as\,e^{b(s)t} + \frac{2}{(8\pi)^2 s} \int_{-\infty}^{0} \int_{-\infty}^{0} dt'\,dt'' \\ \times A(s, t')A(s, t'')T(t, t', t'') \tag{42}$$

where the assumption is made that at high energies the elastic amplitude is essentially imaginary.

Defining $G_s(t) = A(s, t)/(as)$, we have

$$G_s(t) = e^{bt} + \frac{\lambda}{4\pi} \int_{-\infty}^{0} \int_{-\infty}^{0} G_s(t')G_s(t'')T(t, t', t'')b\,dt'\,dt'' \tag{43}$$

with

$$\lambda = \frac{a}{8\pi b}$$

The iterative solution in terms of a power series in λ is

$$G_s(t) = \sum_{n=0}^{\infty} \lambda^n G_n(t) \qquad G_0(t) = e^{bt} \tag{44}$$

Then we have the recurrence relation

$$G_{n+1}(t) = \frac{1}{4\pi} \sum_{m=0}^{n} \int_{-\infty}^{0} \int_{-\infty}^{0} G_m(t')G_{n-m}(t'')\,bT(t, t', t'') \tag{45}$$

which is satisfied if we make the ansatz

$$G_n(t) = \frac{c_n}{n+1}\,e^{bt/n+1} \tag{46}$$

where the c_n satisfy the relation

$$c_{n+1} = \tfrac{1}{4} \sum_{m=0}^{n} c_m c_{n-m} \qquad c_0 = 1 \tag{47}$$

This follows from the fact that

$$\int_{-\infty}^{0} \int_{-\infty}^{0} G_m(t')G_{n-m}(t'')\,bT(t, t', t'')dt'\,dt'' \\ = \frac{\pi}{n+2}\,e^{bt/n+2}c_m c_{n-m} \tag{48}$$

Equation (47) is the relation we would obtain when the quadratic equation

$$c = 1 + \frac{\lambda}{4} c^2 \tag{49}$$

is solved by iteration with

$$c = \sum_{n=0}^{\infty} \lambda^n c_n \tag{50}$$

We obtain

$$c_n = \frac{1 \cdot 3 \cdot 5 \ldots 2n - 1}{4 \cdot 6 \cdot 8 \ldots 2n + 2} = \frac{(2n)!}{4^n (n!)^2 (n+1)} \qquad n \geq 1 \tag{51}$$

For $n \gg 1$, we can use Stirling's formula to get

$$c_n \sim \frac{1}{\pi^{1/2}} \frac{1}{n^{3/2}} \tag{52}$$

Subsituting this in the equation for $A(s, t)$, we obtain for large n

$$
\begin{aligned}
\frac{A(s, t)}{s} &\simeq \frac{a}{\lambda \pi^{1/2}} \sum_{n=0}^{\infty} \frac{1}{n^{5/2}} e^{-n\rho + (bt)/n} \\
&\simeq \frac{a}{\lambda \pi^{1/2}} \int_0^{\infty} \frac{dn}{n^{5/2}} e^{-n\rho + (bt)/n} \\
&= \frac{8\pi}{-t} \left[\frac{1}{2(-bt)^{1/2}} + \rho^{1/2} \right] e^{-2(-b\rho t)^{1/2}}
\end{aligned}
\tag{53}
$$

where

$$\rho = \log \frac{1}{\lambda} = \log \frac{8\pi b}{a} \tag{54}$$

For $|t|$ not too small, we obtain from equation (53)

$$
\begin{aligned}
S = \frac{d \log (A/s)}{d \log s} &= -\frac{\dot{\alpha}}{2b} - \frac{2\dot{\alpha}t(1 + \rho)}{1 + 2(-\rho b t)^{1/2}} \\
&\sim \dot{\alpha} \left[\frac{-(\rho + 1)^2 t}{\rho b} \right]^{1/2} \qquad \text{for } b \to \infty
\end{aligned}
\tag{55}
$$

$$\frac{dS}{d \log s} \sim \frac{\dot{\alpha}(1 + \rho)}{2(-\rho b t)^{1/2}} \qquad \text{as } b \to \infty \tag{56}$$

The corresponding expressions for the Pomeranchuk Regge-pole term $A_0(s, t)$ are

$$S = \dot{\alpha}(t) \qquad \frac{dS}{d \log s} = 0 \tag{57}$$

Now S represents the shrinking of the diffraction pattern, as a function of energy. From equation (55) we see that it is expected to decrease with energy and to increase with t less than the linear dependence on t assumed for $\alpha(t)$. The variation of the shrinking with energy is given by equation (56), whereas it is zero for the corresponding Regge pole. Further, we see that

$$\frac{dS^2}{da} \sim \frac{(1 - \rho^2)\dot{\alpha}^2 t}{\rho ba} \tag{58}$$

The sign of this derivative and hence the dependence of the shrinking on the inelastic cross section of the process considered depends on the value of ρ. At the machine energies available at present, $\rho < 1$, and therefore we expect $(dS^2/da) > 0$. This means that at these energies (up to 30 BeV) the shrinking would be smaller for processes with smaller cross sections, i.e., smaller for πN- than for NN-scattering.

REFERENCES

1. S. Mandelstam, *Phys. Rev.* **112**:1344 (1958).
2. G.F. Chew and S. Frautschi, *Phys. Rev.* **123**:1478 (1961).
3. S.K. Srinivasan and K. Venkatesan, *Nuovo Cimento* **30**: 151, 163 (1963).
4. D. Amati, S. Fubini, A. Stanghellini, and M. Tonin, *Nuovo Cimento* **22**: 569 (1961).
5. L. Bertocchi, S. Fubini, and M. Tonin, *Nuovo Cimento* **25**: 626 (1962).
6. D. Amati, A. Stanghellini, and S. Fubini, *Nuovo Cimento* **26**: 896 (1962).
7. D. Amati, M. Cini, and A. Stanghellini, *Nuovo Cimento* **30**: 193 (1963).
8. L.S. Liu and K. Tanaka, *Phys. Rev.* **129**: 1876 (1963).
9. S. Mandelstam, *Nuovo Cimento* **30**: 1127 (1963).

Regge Poles in Weak Interactions and in Form Factors

K. RAMAN

MATSCIENCE
Madras, India

In this paper I shall present a summary of some recent work on three topics:

1. Regge poles in weak interactions.
2. Asymptotic behavior of scattering amplitudes and subtractions in form factors.
3. Regge poles in weak and electromagnetic form factors.

1. REGGE POLES IN WEAK INTERACTIONS

The main problem in weak interactions is to formulate a dynamic theory that goes beyond the lowest order of perturbation theory.

The work we shall now discuss is the beginning of an analysis of weak reactions using the S-matrix approach. This approach has yielded useful results in a theory of strong interactions, and we believe that it will be a useful approach in weak interactions, also.

The first problem in an S-matrix approach is to analyze the kinematics, which is more complicated because of the unequal masses and parity nonconservation.

One of the kinematically simpler problems has been worked out and will be now discussed, namely, weak reactions involving two spin-zero bosons and two spin-half fermions.[1,2] This has been applied to neutrino reactions on spinless targets and to meson production in neutrino-electron collisions. More complex processes are being worked out.

Some of the main difficulties that arise in an S-matrix approach to weak interactions are the following:

First, any complete theory of weak interactions must include the effects of the strong and electromagnetic interactions to all orders. These may enter in a nontrivial way, e.g., they may decide whether subtractions are necessary or not, and whether the higher-order effects in the weak interaction can be taken into account without leading to divergences.

Second, the zero mass of the neutrino causes difficulties when we consider leptonic weak interactions—both because of the accumulation of thresholds (corresponding to the creation of additional $\nu\bar{\nu}$ pairs) and because of the infrared divergence at the lower limit of some of the dispersion integrals (arising from the infinite range of the interaction). This infrared divergence is probably not a serious difficulty, since physical results do not seem to depend on whether m_ν is exactly zero or only very small, and therefore it should be possible to write the dispersion integrals with a finite neutrino mass and take the limit $m_\nu \to 0$ in the results. Further, it is likely (in contrast to electrodynamics) that the exchange of zero-mass intermediate states is responsible for only a small part of the weak interaction, and that the major part comes from the exchange of massive vector bosons.

The accumulation of thresholds makes it necessary to construct an approximation scheme that does not depend on the distance of singularities from the region of interest.

Recent work suggests that Regge poles (and branch points) are useful and perhaps essential in a description of strong interactions. That they may play a useful role in a theory of weak interactions is suggested by the picture of Regge poles and branch points as emerging from a summation of Feynman diagrams[3] and also by the relation between the theory of weak interactions and the theory of composite particles[4] and the fact that Regge poles provide a starting point for a unified description of high-energy scattering and low-energy composite states.

In the problem worked out, namely, weak reactions with two spin-zero bosons and two spin-half fermions, most of the complication is, of course, algebraic. The details will not be discussed here. The partial-wave analysis is made for the direct process and the crossed process, and is continued to complex J. The unitarity condition is formulated for both processes. For neutrino reactions of the form $\nu + \pi \to l + \pi$ and $\nu + \pi \to l + K$, the most appropriate formulation of the unitarity condition is in terms of intermediate states in the crossed process. The amplitude is written, in terms of the crossed process, in a Regge–

Khuri representation. As a first approximation, the higher-order effects of the weak interaction are assumed to be represented by a vector-boson Regge pole, and the equations are written to retain only strongly interacting (two-meson) intermediate states. Solutions are obtained in the effective-range approximation, and a scheme for obtaining the complete solution is formulated. The complete coupled problem is being considered.

The next step is to consider purely leptonic weak reactions. This is now being done.

2. ASYMPTOTIC BEHAVIOR OF SCATTERING AMPLITUDES AND SUBTRACTIONS IN FORM FACTORS

One of the crucial questions about form factors is their behavior at large values of their argument. This will decide whether any subtractions are required in the dispersion relations for these form factors. As is well known, the existence of subtractions is related to the question of whether the particle involved is elementary or composite. For instance, the asymptotic limit of a form factor may be related to the vertex renormalization constant Z_1 in the usual field theory (for example, see Gell-Mann and Zachariasen[5]). If we postulate that all particles are equally elementary or composite, then the most natural assumption would be that there should be no subtractions in the dispersion relation for any vertex function, or that all form factors go to zero at infinite values of their argument. In a field theory, this would imply that $Z_2 \rightarrow 0$, which may be interpreted as meaning that no particle should be called "elementary."

We shall assume that in fact no subtractions are required for vertex functions, and shall discuss the connection between this no-subtraction principle and the asymptotic behavior of scattering amplitudes.

We first ask what forms of asymptotic behavior for elastic scattering amplitudes are consistent with no subtractions for form factors when the latter are assumed to be dominated by the elastic (two-particle) intermediate states. We shall later see how the argument may be extended to include inelastic intermediate states, if certain assumptions can be made about the behavior of multiparticle scattering amplitudes.

Consider the pion form factor $F(s)$; assume that it satisfies a dispersion relation

$$F(s) = \frac{1}{\pi} \int_{s_0}^{\infty} ds' \frac{\operatorname{Im} F(s')}{s' - s - i\epsilon} \tag{1}$$

If we assume that the two-pion intermediate states dominate the form factors, we have the relation

$$\operatorname{Im} F(s) = F^*(s) T_{\pi\pi}^{J=1}(s) \tag{2}$$

where $T_{\pi\pi}^{J=1}(s)$ is the pion–pion $J = 1$ partial-wave scattering amplitude.

Equation (1) now becomes

$$F(s) = \frac{1}{\pi} \int_{s_0}^{\infty} \frac{F^*(s') T_{\pi\pi}^{J=1}(s')}{(s' - s - i\epsilon)} ds' \tag{3}$$

Equations (2) and (3) imply a restriction on the high-energy behavior of the partial-wave scattering amplitude $T_{\pi\pi}^{J=1}(s)$.

The high-energy behavior of the dispersion integrals is given by the following theorem:[6]
If

$$f(s) \sim s^\alpha \qquad s \longrightarrow \infty$$

then

$$I(s) \equiv \frac{1}{\pi} \int_{s_0}^{\infty} \frac{f(s')}{s' - s} ds' \sim \frac{-(-s)^\alpha}{\sin(\pi\alpha)} \qquad (-1 < \alpha < 0) \tag{4a}$$

$$\sim \text{constant} \qquad (\alpha \leqq -1) \tag{4b}$$

Equations (2), (3), and (4), together with the restriction

$$0 \leqq |T_{\pi\pi}^{(1)}(s)| \leqq 1 \tag{5}$$

which follows from unitarity, lead to the following results:

A. If

$$\operatorname{Im} F(s) \sim s^\alpha \qquad (\alpha \leqq -1) \qquad \text{when } s \longrightarrow \infty$$

then

$$\operatorname{Re} F(s) \sim \frac{1}{s} \qquad\qquad\qquad \text{when } s \longrightarrow \infty \tag{6}$$

B. If

$$\operatorname{Im} F(s) \sim s^\alpha \qquad (-1 < \alpha < 0) \qquad \text{when } s \longrightarrow \infty$$

then

$$\operatorname{Re} F(s) \sim s^\alpha \qquad\qquad\qquad \text{when } s \longrightarrow \infty \tag{7}$$

Equation (2) implies the following: If alternative (6) is true, then (2) implies that

$$T_{\pi\pi}^{(1)}(s) \sim s^{-\gamma_1} \qquad\qquad \text{when } s \longrightarrow \infty \tag{8}$$

where

$$\gamma_1 = -(1 + \alpha) > 0 \tag{8a}$$

On the other hand, when (7) holds, (2) implies that

$$T^{(1)}_{\pi\pi}(s) \sim \text{constant} \qquad\qquad \text{when } s \longrightarrow \infty \qquad (9)$$

We now ask what forms of the elastic amplitude $T(s,t)$ would lead to behavior (8) or (9) for the $J = 1$ partial-wave amplitude $T^{(1)}(s)$.

For this, we consider the behavior of the total amplitude predicted by different models, and see what is that predicted for the partial-wave amplitude in each case. The examples we consider will be particular cases of the general form

$$T(s,t) \sim C(t)s^{\alpha(t)}(\log s)^{\beta(t)} \qquad \text{when } s \longrightarrow \infty \text{ for fixed } t \qquad (10)$$

Single Vacuum Pole

If a single vacuum pole (or the Pomeranchuk trajectory) $\alpha_P(t)$ with $\alpha_P(0) = 1$ dominates the high-energy behavior, then the scattering amplitude behaves as

$$T(s,t) \sim iC(t)s^{\alpha_P(t)} \qquad \text{as } s \longrightarrow \infty \text{ for fixed } t \qquad (11)$$

where $\alpha_P(t)$ is real and less than one for $t < 0$, and

$$\alpha_P(t) = 1 \qquad \text{for } t = 0 \qquad (11a)$$

The behavior of $\alpha_P(t)$ for $t \longrightarrow \infty$ and the form of $C(t)$ are unknown. We shall assume that $\alpha_P(t)$ is almost linear in the neighborhood of $t = 0$ and that it approaches some negative integer $-n$ as $t \longrightarrow -\infty$, where $n > 0$, for instance:

$$\alpha_P(t) \sim -n + \frac{\text{constant}}{(t - t_0)A} \qquad \text{as } t \longrightarrow \infty \text{ with } n > 0, A > 0 \qquad (11b)$$

Regarding the behavior of $C(t)$, we examine different possibilities:

(i) $\qquad C(t) \sim |t| \quad \text{for } t \approx 0 \qquad C(t) \sim \dfrac{1}{t} \quad \text{for } t \longrightarrow -\infty$

which is suggested by potential scattering. (Potential scattering suggests that $C(t) \sim |t|$ for t near threshold.)

(ii) $\quad C(t) \sim |t|^{-\gamma} \quad \text{for } t \approx 0 \qquad C(t) \sim |t|^{-\gamma} \quad \text{for } t \longrightarrow -\infty$

(iii) $\qquad C(t) \sim [C(0) + tC'(0) + \cdots] \quad \text{for } t \approx 0$

$$C(t) \sim |t|^{-\gamma} \quad \text{for } t \longrightarrow -\infty$$

(iv) $\qquad C(t) \sim [C(0) + tC'(0) + \cdots] \quad \text{for } t \approx 0$

$$C(t) \sim e^{\gamma t} \quad \text{for } t \longrightarrow -\infty$$

(v) $\quad C(t) \sim |t|^{-\gamma} \quad \text{for } t \approx 0 \qquad C(t) \sim e^{\gamma t} \quad \text{for } t \longrightarrow -\infty \qquad (11c)$

Here,

$$C > 0 \qquad \bar{\gamma} > 0 \qquad 0 < \gamma < 1 \qquad\qquad (11d)$$

The results obtained are that neither (8) nor (9) can be obtained *unless* $\alpha'_P(0) = 0$, assuming that $\alpha_P(-\infty) < 0$. If $\alpha_P(-\infty)$ is positive, it is possible to obtain the behavior (9) even if $\alpha'_P(0) \neq 0$. However, we believe that it is very unlikely that $\alpha_P(-\infty) < 0$; it is more probable that $\alpha_P(-\infty) = -1$.

Thus, if the form factors are dominated by elastic intermediate states, a simple single-vacuum-pole Regge behavior is not consistent with a no-subtraction philosophy for form factors, unless $\alpha'_P(0) = 0$.

Recently, Sugawara has shown that if $\alpha'_P(0) = 0$, then unitarity and analyticity imply that $\alpha_P(t) = 1$ for all t. This would correspond to a vacuum trajectory that is a fixed pole, or an "elementary" pole, in angular momentum.

We thus have the somewhat strange conclusion that a no-subtraction principle for vertex functions (which may perhaps be taken as equivalent to the statement that no particles are elementary) is consistent with a single-vacuum-pole behavior of the elastic amplitude only if this pole is a fixed pole. (Such a fixed pole would give a nonshrinking diffraction peak.)

Oehme's argument[8] that no fixed pole with $J = 1$ can exist is not consistent with a no-subtraction philosophy for form factors.

Another context in which it has been found that the dominance of a single vacuum pole is not consistent with an unsubtracted dispersion relation for form factors is the Goldberger–Treiman relation; the assumption that the $N\bar{N}$-scattering amplitude was dominated by the vacuum pole would not permit an unsubtracted dispersion relation to be written for the pion decay form factor.[9] They suggested that one should therefore write a dispersion relation with a subtraction.

We believe that this is an incorrect conclusion and that the correct answer is that one should still write an unsubtracted dispersion relation and that the $N\bar{N}$-elastic amplitude is *not* dominated by the vacuum trajectory.

We shall interpret these results as indicating that the high-energy behavior of an elastic amplitude is, in general, more complicated than a single-Regge-pole behavior. We now briefly examine more complicated forms of high-energy behavior.

Two-Vacuum-Pole Behavior

Frye[10] has given a model with two vacuum trajectories which predicts the following behavior for the partial-wave amplitudes:

$$\text{Im } T^{(l)}(s) \sim \text{constant} \qquad s \to \infty$$
$$\text{Re } T^{(l)}(s) \to 0 \qquad s \to +\infty \tag{12}$$

This is consistent with (9), and therefore with no subtractions for form factors.

Optical Sphere Behavior

The classical picture of high-energy scattering as diffraction scattering by an optical sphere[11] predicts a behavior

$$T(s, t) \sim isf(t) \tag{13}$$

where $f(t)$ is an unknown function. This is found to be consistent with (9) if

$$f(t) > \frac{1}{t} \qquad \text{for small } t, \text{ say, } 0 > t > -t_1 \tag{14}$$

and

$$f(t) < \frac{1}{t^{1+\epsilon}} \qquad \text{as } t \to -\infty, \text{ where } \epsilon > 0$$

An interesting possibility is a superposition of (11) and (13):

$$T(s, t) \sim iC(t)s^{\alpha_p(t)} + isf(t) \tag{15}$$

(for example, see Freund and Oehme[12]). This will be consistent with (9) if $f(t)$ has the form (14).

The Multiperipheral Model[13]

The multipheripheral model to first order gives a behavior of the form (11). The model also predicts that $\alpha'(t) > 0$ for $t < 0$; and therefore this behavior is not consistent with no subtractions for the form factors.

The branch points obtained on iterating the elastic terms probably lie in an unphysical sheet (see Polkinghorne[14]); we do not know what is the prediction for the complete amplitude in the physical sheet.

Fixed Branch Points in Angular Momentum

These can arise if the interaction is singular. For instance, Freund and Oehme have considered a behavior of the form

$$T(s, t) \sim iC(t)s^{\alpha(t)} + b(t)\frac{s}{(\log s)^{3/2}} \tag{16}$$

arising from such branch points together with a vacuum pole.[15]

Such a model can be consistent with (9), provided the residue at the branch points has a suitable behavior. For equation (16) the requirement is that

$$b(t) > \frac{1}{t} \qquad \text{for small } t, \text{ say } 0 > t > -t_1$$

and

$$b(t) \sim \frac{(\log|t|)^{1/2}}{t} \qquad \text{as } t \to -\infty \tag{17}$$

Moving Branch Points in Angular Momentum

These are being investigated at present; nothing about them will be mentioned now.

When inelastic intermediate states I_{inel} are also included in the unitarity condition for the form factor, then it is found that if the behavior of the scattering amplitude $I_{\text{inel}} \to I_{\text{inel}}$ is of the same form as for elastic amplitudes (as would be possible if the inelastic intermediate states could be approximated by effective two-particle states, e.g., when we have isobars), then the coupled equations for the form factors lead to a condition of the form (9), which may hold under the same conditions discussed above for the elastic amplitude.

3. REGGE POLES IN WEAK AND ELECTROMAGNETIC FORM FACTORS

We will give a very brief account of some work being done on form factors.

For the pion form factors, the integral equation is first written in the approximation of elastic unitarity, where the low-energy pion–pion scattering amplitude is approximated using a Regge–Khuri representation and the high-energy part is taken to be of the form $[A + (B/\log s)]$. This integral equation is being solved, first by taking an effective-range approximation for the low-energy scattering amplitude. Better solutions will also be obtained.

Next, the integral equation is written with inelastic unitarity, including $K\bar{K}$ states, so that coupled equations are obtained for the π and K form factors. These are being solved with similar assumptions.

An idea of the behavior of the form factors may be obtained by first approximating the form factors by the partial-wave projection of the "full contribution" of the appropriate Regge poles (in the sense of Khuri[16]).

This can be done only for values of s in $F(s)$ that are above threshold, because only then is the Regge–Khuri representation valid. The usual values of s that are of interest are the values $s < 0$. For these, an appropriate form that may be continued below threshold is obtained using a representation recently proposed by Eden,[17] in which Legendre transforms with respect to variables other than $\cos \theta$ are considered.

The difficulty now is that the partial-wave projection is not easy to obtain. However, this can be done numerically. This problem is being worked out.

This difficulty does not, of course, arise when we take the Regge–Khuri representation for the elastic amplitude in the dispersion integral, because then only values of the amplitude above threshold are required.

Both weak and electromagnetic form factors are being considered in all the above.

REFERENCES

1. K. Raman, Ph.D. thesis.
2. K. Raman, *Nuovo Cimento* (to be published).
3. J.C. Polkinghorne, *J. Math. Phys.* **4**: 503 (1963).
4. K. Raman, *Proc. First Anniversary Symposium*, Matscience, Madras, India, January 1963.
5. M. Gell-Mann and F. Zachariasen, *Phys. Rev.* **123**: 1065 (1961).
6. S. Mandelstam, *Ann. Phys.* **21**: 302 (1963).
7. H. Sugawara, *Progr. Theoret. Phys.* **30**: 404 (1963).
8. R. Oehme, *Phys. Rev.* **130**: 424 (1963).
9. G. Barrett and C. Barton, *Nuovo Cimento* **29**: 703 (1963).
10. G. Frye, *Phys. Rev. Letters* **8**: 494 (1962); *Phys. Rev.* **129**: 1453 (1963).
11. M. Froissart, in: "Theoretical Physics," Lectures at Trieste, IAEA (1962).
12. P.G.O. Freund and R. Oehme, *Phys. Rev. Letters* **10**: 199 (1963).
13. Bertocchi, Fubini, and Tonin: *Nuovo Cimento* **25**: 626 (1963).
14. J.C. Polkinghorne, Cambridge preprint.
15. P.G.O. Freund and R. Oehme, *Phys. Rev. Letters* **10**: 450 (1963).
16. N.N. Khuri, *Phys. Rev.* **130**: 429 (1963).
17. R.J. Eden, CERN preprint (1963).

Effective-Range Approximation Based on Regge Poles

B. M. UDGAONKAR

TATA INSTITUTE OF FUNDAMENTAL RESEARCH
Bombay, India

During the last year, a new kind of effective-range approximation[1,2,3] based on singularities in the complex angular momentum plane has given interesting results in several problems concerning low-energy scattering. The expression "effective-range approximation" is used here in the customary sense for the S-matrix theory, namely, an approximation based on the idea that in a limited region of energies the scattering amplitude may be reasonably well parameterized in terms of a few nearby singularities. It is well known that such an approximation in terms of singularities in the complex energy or momentum plane provides a very good description of low-energy scattering of, for example, the n-p system. We shall see in what follows how singularities in the complex angular momentum plane can also be used for the same purpose.

In order to bring out the relationship of this use of nearby singularities to the usual effective-range expansion, let us write the S-matrix element for np-scattering in the ^3S state:

$$
\begin{aligned}
S \equiv e^{2i\delta} &= \frac{\cot \delta + i}{\cot \delta - i} = \frac{-a + (r_0/2)(k^2 + a^2) + ik}{-a + (r_0/2)(k^2 + a^2) - ik} \\
&= \frac{(k + K_1)(k + K_2)}{(k - K_1)(k - K_2)}
\end{aligned}
\tag{1}
$$

where $K_1 = ia$ is the bound-state pole on the imaginary axis in the k-plane, and $K_2 = i[(2/r_0) - a]$ is another pole on the imaginary axis, which may be regarded as effectively replacing all the other singularities in the k-plane. Expression (1) also has the correct zeros associated with these poles. We know, of course, that in general the S-matrix has an infinite number of poles and also has cuts along the imaginary axis, and we see from above that what one does in making an effective-range approximation of the usual type is to approximate to it by a couple of poles.

Let us consider some other well-known effective-range formulas— for example, take the Chew–Low formula for the $\frac{3}{2}$–$\frac{3}{2}$ phase shift in πN-scattering:

$$\frac{k^3}{\omega} \cot \delta_{33} = \frac{3}{4 f^2} \left(1 - \frac{\omega}{\omega_r} \right) \tag{2}$$

where $\omega = W - m$, $\omega_r = W_R - m$, W being the total energy in the CM system and W_R the total energy in the CM system at resonance. Equation (2) may be rewritten as

$$
\begin{aligned}
f_{33} &\equiv \frac{e^{i\delta_{33}} \sin \delta_{33}}{k^3} \\
&\approx \gamma_{33} \left[\frac{1}{\omega} - \frac{1}{\omega - \omega_r + i\Gamma_{33}/2} \right]
\end{aligned}
\tag{3}
$$

Here $\gamma_{33} = 4 f^2/3$; $(\Gamma_{33}/2) = (k^3 \gamma_{33} \omega_r/\omega)$. Thus, we may say that the Chew–Low formula is an effective-range formula which expresses the $\frac{3}{2}$–$\frac{3}{2}$ amplitude in terms of two dominant singularities, namely, the crossed nucleon pole and the pole corresponding to the resonance itself. Its success means that once these two singularities are taken into account, no important singularities are being missed.

Consider now another example where the usual effective-range formula is not so successful, the Chew–Low formula for the $J = \frac{1}{2}^+$, $T = \frac{1}{2}$ p-wave phase-shift:

$$f_{11} \equiv \frac{e^{i\delta_{11}} \sin \delta_{11}}{k^3} = \frac{-8 f^2/3\omega}{1 + (\omega/\omega_r)} = \frac{-8 f^2}{3} \left[\frac{1}{\omega} - \frac{1}{\omega_r + \omega} \right] \tag{4}$$

Equation (4) expresses the amplitude in terms of the two nucleon poles—a direct nucleon pole of residue $-3 f^2$, and a crossed nucleon "pole" of residue $+\frac{1}{3} f^2$ and a pole at $-\omega_r$ representing the fuzzy cut due to the crossed $\frac{3}{2}$–$\frac{3}{2}$ resonance.

As already mentioned, equation (4) is known to be a poor representation of the amplitude f_{11}. Its inadequacy means that some important singularities of this amplitude have been left out, e.g., the contribution of ρ. An effective-range formula for this amplitude based on energy-plane singularities must therefore be much more complicated. We shall see later that one can get a rather simple effective-range formula for this amplitude in terms of the singularities in the complex angular momentum plane.

Before seeing how this can be done, we have to consider an important modification of the original Regge representation which was proposed by Khuri.[1] Without such a modification, the Regge representation could not be used for an effective-range approximation. To see this, let us write the Regge representation as

$$A(s,z) = -\pi \sum_{i=1}^{N} [2\alpha_i(s) + 1] \frac{\beta_i(s) P_{\alpha_i(s)}(-z)}{\sin \pi \alpha_i(s)}$$
$$+ \frac{i}{2} \int_{-1/2-i\infty}^{-1/2+i\infty} (2l + 1) \frac{P_l(-z)}{\sin \pi l} A(l, s) dl \tag{5}$$

This representation is valid, for example, for potentials which are superpositions of Yukawa potentials, that is,

$$rV(r) = \int_{m_0}^{\infty} \sigma(\mu) e^{-\mu r} d\mu \tag{6}$$

The summation in equation (5) goes over all the resonances and bound states of the problem. Now, the number of known resonances and bound states in any scattering problem—πN, NN, KN, etc.—is rather small, and so it is very tempting to use representation (5), with only a few pole contributions corresponding to these known bound states and resonances, in order to approximate the scattering amplitude. Representation (5), without the background term which we would like to omit in such an approximation, has, unfortunately, two undesirable features. First, the individual terms in the summation, and therefore the sum itself, have a wrong cut in the z-plane, starting at $z = 1$ instead of at $z = 1 + (m_0^2/2s)$. Second, if one projects out a partial wave from one of these Regge pole terms in equation (5), the correct threshold behavior is not obtained. Obviously, an expression which is to be used for an effective-range approximation at low energies must at least have the correct threshold behavior, and the correct cut.

These deficiencies of the original Regge representation were removed by Khuri, by decomposing the background integral into two parts, one of which is added to the usual Regge pole term to give "the full contribution" of the Regge pole. For this, Khuri uses the following representation of the Legendre functions, valid for $-1 < \mathrm{Re}\, l < 0$ and for $-1 < z < 1$:

$$\frac{(2l + 1)\pi P_l(z)}{\sin \pi l} = \frac{-1}{\sqrt{2}} \int_{-\infty}^{\infty} \frac{e^{(l+\frac{1}{2})x} \sinh x}{(\cosh x + z)^{3/2}} dx \tag{7}$$

Substituting this in equation (5) and interchanging the order of integration, we get

$$A(s, z) =$$

$$\frac{1}{\sqrt{2}} \int_{-\infty}^{\infty} \frac{B(x, s)\,\sinh x}{(\cosh x - z)^{3/2}}\,dx - \pi \sum_{i=1}^{N} \frac{(2\alpha_i(s) + 1)\beta_i(s)P_{\alpha_i(s)}(-z)}{\sin \pi \alpha_i(s)}$$

$$(8)$$

where

$$B(x, s) = \frac{1}{2\pi i} \int_{-\frac{1}{2}-i\infty}^{-\frac{1}{2}+i\infty} dl\, e^{(l+\frac{1}{2})x} A(l, s) \tag{9}$$

We now write

$$\int_{-\infty}^{\infty} \frac{B(x, s)\,\sinh x}{(\cosh x - z)^{3/2}}\,dx = \int_{-\infty}^{\xi} \frac{B(x, s)\,\sinh x\,dx}{(\cosh x - z)^{3/2}} + \int_{\xi}^{\infty} \frac{B(x, s)\,\sinh x\,dx}{(\cosh x - z)^{3/2}} \tag{10}$$

where $\cosh \xi \equiv z_0 = 1 + (m_0^2/2s)$. In the first of these integrals, on the right-hand side, where $x < \xi$, we can close the contour in equation (9) on the right and get

$$B(x, s) = - \sum_{i=1}^{N} \beta_{i(s)}e^{(l+\frac{1}{2})x} \qquad \text{for } x < \xi \tag{11}$$

Equation (8) may now be rewritten as

$$A(s, z) = \sum_{i=1}^{N} R(s, z; \alpha_i) + \frac{1}{\sqrt{2}} \int_{\xi}^{\infty} \frac{B(x, s)\,\sinh x\,dx}{(\cosh x - z)^{3/2}} \tag{12}$$

where

$$R(s, z; \alpha_i) \equiv -\beta_i(s)\left[\frac{\pi(2\alpha_i(s) + 1)P_{\alpha_i(s)}(-z)}{\sin \pi\, \alpha_i(s)} \right. $$
$$\left. + \frac{1}{\sqrt{2}} \int_{-\infty}^{\xi} \frac{e^{(\alpha_i+\frac{1}{2})x}\,\sinh x\,dx}{(\cosh x - z)^{3/2}}\right] \tag{13}$$

which is valid for $\operatorname{Re} \alpha_i > -\frac{1}{2}$.

So far we have restricted z to physical values. We can now use equations (12) and (13) to make an analytic continuation in z, and see that the new background integral in equation (12) and therefore $\sum_{i=1}^{N} R(s, z; \alpha_i)$ has the correct cut in z. Khuri, in fact, explicitly shows that $R(s, z; \alpha_i)$ has the correct cut from z_0 to ∞. Also, if one projects the l-th partial wave from $R(S, Z; \alpha_i)$, one gets

$$r(s, l; \alpha_i) = -\beta_i(s)e^{-(l-\alpha_i)\xi\,(\alpha_i-l)} \tag{14}$$

which has the correct threshold behavior characteristic of an l-th partial wave. To see this, we have only to note that, as shown by Barut and Zwanziger

$$\beta_i(s) \sim s^{\alpha_i(0)} \qquad \text{for small } s \qquad (15)$$

and

$$e^{-(l-\alpha_i)\xi} \sim s^{l-\alpha_i(0)} \qquad \text{for small } s$$

so that

$$r(s, l; \alpha_i) \sim s^l \qquad \text{for small } s$$

It is therefore very tempting to identify each term $R(s, z; \alpha_i)$ in summation (12) as the "full" contribution to $A(s, z)$ of a Regge pole in the right half-plane. This identification is made even more meaningful by the fact that for potentials for which $A(l, s)$ is meromorphic over the entire l-plane, Khuri has been able to show that the new background integral in equation (12) is just a sum over the "full contributions" of the poles in the left-hand half-plane.

Representation (12) can thus be made the basis of a new effective-range approximation, where one may retain only a few right-hand poles in the summation, and neglect the new background integral. In order that the latter be justified the range of energies has to be such that $\xi > 1$.

An application we shall consider in some detail is the determination of the $J = \frac{1}{2}^+$, $T = \frac{1}{2} \pi N$ phase shift in terms of the nucleon Regge pole. We shall not go through the algebraic complications arising from the fact that we are here dealing with a particle with spin, but merely quote the resulting expression for the contribution of the nucleon Regge pole to this partial wave. It is

$$a_+(\tfrac{1}{2}, W) \equiv \frac{e^{i\delta_{11}} \sin \delta_{11}}{k} = \frac{\beta(W)}{2[\alpha(W) - \frac{1}{2}]} [e^{[\alpha(W) - \frac{1}{2}]\xi_1} + e^{[\alpha(W) - \frac{1}{2}]\xi_2}]$$

$$(16)$$

Here $\cosh \xi_1 = 1 + (2/k^2)$ and $-\cosh \xi_2 = -[(W^2 - m^2 - 2)/2k^2 - 1]$, and these two quantities represent the beginning of the right-hand and the left-hand cuts in z, respectively. If the nucleon is the only $T = \frac{1}{2}$ trajectory of even J-parity which lies close to the physical region at low energies, then equation (16) should be a good approximation.

Before we can use this expression we must know the functions $\alpha(W)$ and $\beta(W)$, which represent the trajectory of the nucleon pole and its residue, respectively. These functions are not yet known, and so we have to make some approximations. We know that $\alpha(m) = \frac{1}{2}$, and in the region between the nucleon and the N^{***} it is not unreasonable to assume that Re $\alpha(W)$ may be approximated by a straight line

$$\text{Re } \alpha(W) = \tfrac{1}{2} + \epsilon(W - m)$$

For the slope of the trajectory, we shall take the value determined by the position of the N^{***}, namely, $\epsilon = 0.4$. As for $\beta(W)$, we know that $\beta(W)$ has the following threshold behavior:

$$\beta(W) \sim \left(\frac{k^2}{s_0}\right)^{\alpha_0 + 1/2} \qquad \alpha_0 = \alpha(W)|_{k^2=0}$$

Hence, the product

$$\beta(W) \exp\left[(\alpha + \tfrac{1}{2})\xi\right]$$

would be slowly varying near the threshold. We shall assume it to be a real constant C in a limited-energy region. Expression (16) then becomes

$$a_+(\tfrac{1}{2}, W) = \frac{C}{2\epsilon}\left[\frac{\{\exp(-\xi_1) + \exp[(\alpha + \tfrac{1}{2})(\xi_2 - \xi_1) - \xi_2]\}}{W - m}\right]$$

$$(17)$$

The constant C/ϵ can now be determined by extrapolation to the nucleon pole and by demanding that our expression reduce to the usual perturbation theory pole term, namely,

$$a_+(\tfrac{1}{2}, W) \xrightarrow[W \to m]{} \frac{3f^2}{W - m} \tag{18}$$

We then get

$$a_+(\tfrac{1}{2}, W) =$$
$$-\tfrac{3}{2}f^2\left[\frac{\exp(-\xi_1) + \exp\{[1 + \epsilon(W - m)](\xi_2 - \xi_1) - \xi_2\}}{W - m}\right] \tag{19}$$

This expression can be easily seen to have the correct threshold behavior of a p-wave amplitude, i.e., $a_+(\tfrac{1}{2}, W) \sim k^2$ as $k^2 \to 0$.

Expression (19) has no arbitrary constants, and one finds that it gives reasonably good agreement with the experimental data. Previous effective-range calculations, on the other hand, had all tended to give a phase-shift considerably larger in magnitude than that observed. For $k^2 \ll 1$, our expression gives $a_+ = -0.96 f^2 k^2$, while the perturbation theory pole gives $a_+ = -3 f^2 k^2$. Thus, the effect of reggeizing the nucleon has been to cut down its contribution even at threshold, compared to what it is when it is treated as an ordinary particle.

Bose and Der-Sarkissian[4] have applied this method to np-scattering, treating the deuteron as a Regge pole. In this case, there is not much difference between the ordinary effective-range approximation and the Regge-pole approximation. This is probably because the deuteron pole lies so close to the physical region. The method has also been applied to the $(\tfrac{3}{2}\text{-}\tfrac{3}{2})$ phase-shift in πN-scattering by Biswas and

Bose,[5] again with good results. Biswas and Bose[6] have also considered an application to a multichannel problem, namely, the contribution of Y_0^* to $\overline{K}N$-scattering, and the results look quite promising.

REFERENCES

1. N.N. Khuri, *Phys. Rev.* **130**: 429 (1963).
2. N.N. Khuri and B.M. Udgaonkar, *Phys. Rev. Letters* **10**: 172 (1963).
3. B.M. Udgaonkar, in: *Strong Interactions and High Energy Physics*, Oliver and Boyd, London, and Plenum Press, New York (1964).
4. S.K. Bose and M. Der-Sarkissian, *Nuovo Cimento* **30**: 878 (1963).
5. S.N. Biswas and S.K. Bose, *Phys. Rev.* **133**: B 789 (1964).
6. S.N. Biswas and S.K. Bose, Private communication.

Some Applications of Separable Potentials in Elementary Particle Physics

A. N. MITRA[*]

UNIVERSITY OF DELHI
Delhi, India

Our group at Delhi University has recently done some work on the application of the so-called separable interactions to certain problems of current interest in elementary particle physics.

Separable potentials were first discovered by Wigner (mostly unpublished), who found that they satisfy the usual requirements of translational and time-reversal invariance. They are also known to produce saturation of nuclear forces. These potentials, which are highly nonlocal, have shapes given by one of the following expressions:

$$\langle \mathbf{r} | V | \mathbf{r}' \rangle = \sum_{l=0}^{\infty} \lambda_l (2l + 1) g_l(r) g_l(r') P_l(\hat{\mathbf{r}} \cdot \hat{\mathbf{r}}') \tag{1}$$

$$\langle \mathbf{p} | V | \mathbf{p}' \rangle = \sum_{l=0}^{\infty} \lambda_l (2l + 1) f_l(p) f_l(p') P_l(\hat{\mathbf{p}} \cdot \hat{\mathbf{p}}') \tag{2}$$

where $f_l(p)$ is just the Hankel transform of $g_l(r)$ (in going from coordinate space to momentum space). These partial-wave expansions suggest that such shapes should be particularly useful for low-energy processes involving very few l-values. Spin-dependent and noncentral forces (tensor, spin-orbit, etc.) in separable form have also been used, and have representations similar to equations (1) or (2).

The first quantitative calculations using separable potentials were made by Yamaguchi,[1] who applied them to the problems of the deuteron structure, n-p scattering, and photodisintegration of the deuteron. The main advantage of separable potentials, as was found by Yamaguchi, is that they render a two-body problem exactly soluble.

---

[*] Department of Physics and Astrophysics.

I shall not discuss the calculations that we have made with separable potentials in the domain of low-energy nuclear physics, such as the detailed fits to the two-body parameters,[2-4] binding energy in nuclear matter, spin-orbit splitting of energies in odd nuclei,[2] energy levels of some odd–odd nuclei,[5,6] etc., since most of these investigations have been published earlier. Instead, this paper will be generally limited to some more recent work on (a) certain formal properties that we have found in separable potentials, and (b) the extension of the concept of separable potentials to the domain of relativistic processes involving elementary particles.

As for the formal properties, I shall first briefly indicate certain results on analyticity properties, double dispersion representation, and Regge behavior of amplitudes constructed from separable interactions.[7,8]

A potential of the form

$$\langle \mathbf{p}|V|\mathbf{p}'\rangle = -\frac{\lambda}{M} \sum_{l=0}^{\infty} (2l+1)v_l(p)v_l(p')P_l(\hat{\mathbf{p}}\cdot\hat{\mathbf{p}}') \tag{3}$$

leads to the elastic scattering amplitude

$$f(s,t) = \sum_{l=0}^{\infty} (2l+1)A_l(s)P_l\left(1 + \frac{t}{2s}\right) \tag{4}$$

where

$$s = k^2 \qquad\qquad t = -2k^2(1 - \cos\theta) \tag{5}$$

$$A_l(s) = \frac{N_l(s)}{D_l(s)} \tag{6}$$

$$N_l(s) = 2\pi^2\lambda v_l^2(\sqrt{s}) \tag{7}$$

$$D_l(s) = 1 - \frac{1}{\pi}\int_0^\infty ds' \frac{\sqrt{s'}\,N_l(s')}{s' - s} \tag{8}$$

Taking a shape given by

$$v_l^2(p) = \frac{2}{p^2}\,Q_l(1 + \tfrac{1}{2}\beta^2 p^{-2}) \tag{9}$$

one obtains a result which corresponds to the first Fredholm order of truncation to the solution for a partial-wave amplitude due to a Yukawa potential of range β^{-1}.*

The derivation of the double dispersion representation for $f(s,t)$ given by equation (4) can be made with the Regge formalism.[8] If we recall the pre-Regge "proofs" of the Mandelstam representation, these

---

* The left-hand cut in the N_l-function given by equations (7) and (9) starts at $K^2 = -\tfrac{1}{4}\beta^2$.

depended on the part played by the two variables s and t. However, the Regge formalism showed that the "proofs" would go through in a less restrictive way (without the assumption of bounded behavior as $t \rightarrow \infty$), with the scheme of the variables s, l. Now, with the separable potentials used here, it looks very natural to use this latter scheme, since it forms the very basis of representation of the interaction. Thus, one has to convert equation (4) to a Watson–Sommerfeld representation and thence to the Regge representation by deforming the contour to pass over the angular momentum poles. It turns out in this theory that there is just one Regge pole, which is also the principal pole. The behavior of its trajectory and the high-energy behavior of the crossed amplitude follow rather closely the Yukawa pattern if the form given by equation (9) is used for the potential. Further, using such a Regge representation,

$$f(s, t) = \frac{8\pi^2 \lambda}{\beta^2 - t} + \tfrac{1}{2}i \int_{l_0 - i\infty}^{l_0 + i\infty} dl \frac{(2l + 1)}{\sin \pi l} B_l(s) P_l \left(-1 - \frac{t}{2s} \right)$$

$$+ (2\alpha + 1) \frac{\beta(s) P_\alpha \left(-1 - \dfrac{t}{2s} \right)}{\sin \pi \alpha} \tag{10}$$

where

$$B_l(s) = A_l(s) - N_l(s) \tag{11}$$

and $\alpha(s)$ is the Regge pole, it is possible to show that $f(s, t)$ satisfies a double dispersion representation with cuts along (i) $s > 0$, $t > 0$ (ii) $t > 0$, $s < -\tfrac{1}{4}\beta^2$. The second cut, which has no analog in Yukawa potential scattering, corresponds to the cut $u > 0$, $t > 0$, where $u + 4s + t = 0$. This is a characteristic field-theoretical effect, and seems to indicate that our separable potential, which could presumably come about through some sort of effective-range approximation to a model field theory, has some built-in residual field-theoretical features. Investigation of such formal analytic properties was motivated by the desire to see if separable interactions do indeed conform to certain standard mathematical disciplines and admit of dispersion-theoretical interpretations, without being mere empirical forms of parameterization for advantage in calculation. We therefore consider it rather satisfying that this is true.

Another formal property of separable potentials concerns the solution of a three-body problem.[9] Let us consider an oversimplified situation where three identical, massive particles (mass μ), without any internal degrees of freedom, are in s-wave interactions in pairs, the potential for one pair (i, j) being

$$\langle \mathbf{P}_i \mathbf{P}_j | V | \mathbf{P}'_i \mathbf{P}'_j \rangle = -\frac{\lambda}{\mu} v(\mathbf{p}_{ij}) v(\mathbf{p}'_{ij}) \delta^3(\mathbf{P}_k - \mathbf{P}'_k) \qquad (12)$$

where

$$2\mathbf{p}_{ij} = \mathbf{P}_i - \mathbf{P}_j \qquad\qquad -\mathbf{P}_k = \mathbf{P}_i + \mathbf{P}_j \qquad (13)$$

and momentum conservation has been taken into account. Then the three-body wave function Ψ satisfies

$$\Delta(E)\Psi = \lambda \sum_{ijk} \int d^3\mathbf{p}'_{ij} v(\mathbf{p}_{ij}) v(\mathbf{p}'_{ij})$$
$$\times \Psi(-\tfrac{1}{2}\mathbf{P}_k + \mathbf{p}'_{ij}, -\tfrac{1}{2}\mathbf{P}_k - \mathbf{p}'_{ij}, \mathbf{P}_k) \qquad (14)$$
$$\Delta(E) = \tfrac{1}{2}(P_1^2 + P_2^2 + P_3^2) - E\mu$$

and one sees immediately that it has the structure

$$\Psi = \Delta^{-1}(E)[v(p_{12})F_3(P_3) + v(p_{23})F_1(P_1) + v(p_{31})F_2(P_2)] \qquad (15)$$

Use of the symmetry requirement leads further to, say,

$$F_1 \equiv F_2 \equiv F_3 = F \qquad (16)$$

where

$$\left[1 - \frac{1}{\pi} \int_0^\infty \frac{N(s')\, ds'}{\sqrt{s}\,(s'^2 + \tfrac{3}{4}P_1^2 - E\mu)} \right] F(P_1)$$
$$= 2\lambda \int d^3\mathbf{q}(P_1^2 + q^2 + \mathbf{P}_1 \cdot \mathbf{q} - E\mu)^{-1} v(\mathbf{q} + \tfrac{1}{2}\mathbf{P}_1) F(q) \qquad (17)$$

Equation (15) has the simple interpretation that each term represents the wave function $[v(p_{ij})]/\Delta(E)$ of a pair—(i, j) multiplied by the relative function $F(P_K)$ between K and (i, j). The "pole" in $F(P_K)$, given by equation (17) shows explicitly how the resonating group structure idea of Wheeler is satisfied in this model.

Such explicit three-particle structures can be conveniently used to study certain formal properties of three-particle systems which are largely based on conjectures. One such property concerns the so-called unitarity condition which was conjectured by Blankenbecler[10] to require that at bound state energies of a three-particle system, the "phase" of the three-particle amplitude should be equal to that of a pair of its members whose relative energy allows them to be in a physical scattering state. This question can be directly tested in terms of equations (15) to (17), and it is found *not* to be the case,[11] essentially because the requirement of symmetrization within a two-body potential model necessitates an *additive* structure for the amplitude, where only one of the terms satisfies the Blankenbecler requirement, but not the two others. Thus, the overall phase of the three-particle amplitude is different from that of the relevant two-particle pair.

Apart from such formal investigations, one can use this formalism to calculate the physical parameters for various phenomena involving low-energy three-particle systems. Even with the inclusion of the appropriate spin and isospin variables, we have calculated (i) the low-energy parameters of n-d scattering,[12] (ii) binding energy of the triton, and (iii) the energy distribution of the odd pion in decay.[13]

In the case of n-d scattering,[12] we have been able to resolve the ambiguity in the two sets of experimental scattering lengths corresponding to $S = \frac{3}{2}$ and $S = \frac{1}{2}$. We find that the set which gives $a_{3/2} = (6.4F \pm 0.3)$ is the correct one, and our value for $a_{3/2}$ agrees very closely with the above value.

For the bound three-body problem, our calculations using full antisymmetrization of the three-nucleon wave function show that the binding energy comes out too large with an *effective* two-body central force in the triplet state. However, one obtains almost the right amount of binding when tensor forces are taken into account in the three-body formalism.[14] The results with Yamaguchi's tensor force parameters[15] show a binding energy of about 9 MeV, as against a value of 13 MeV when only the effective central force parameters are used.[16] It may be emphasized that these values are based on exact (numerical) evaluations of certain integral equations, and not on the usual variational treatments for a three-body problem.

For $K\pi_3$ decay,[13] our result indicates that in order to explain the high-energy bias of the odd pion, one must invoke the hypothesis that the $T = 2$ π-π interaction should be stronger than $T = 0$, in agreement with dispersion-theory findings. Since, on the other hand, there is evidence for stronger $T = 0$ interactions, one must draw the conclusion that the intrinsic structure of the weak interaction plays the dominant role in explaining the anomaly.

To extend these techniques to the relativistic domain, one has to change the dynamic basis from a nonrelativistic Schrödinger equation to one which takes account of relativistic energies. That such extension is possible is made plausible from various approximations in field theory, e.g., the Tamm–Dancoff approximations where, for the π-N wave function, one obtains an equation of the type[17]

$$(E_p + \omega_p - E)\Psi(\mathbf{p}) = \lambda \int d^3\mathbf{p}' \, K(\mathbf{p}, \mathbf{p}')\Psi(\mathbf{p}') \qquad (18)$$

Such a structure can be adapted in our scheme with the assumption of a separable kernel K for each angular momentum state. We have applied this type of model to p-wave pion–pion interaction[18] to produce the ρ-resonance and then used this interaction in pairs to study the structure of ω as a three-pion system—and not merely a $\rho + \pi$ system—through a relativistic Schrödinger equation of the form

$$(\omega_1 + \omega_2 + \omega_3 - E)\Psi = -(V_{12} + V_{23} + V_{31})\Psi \qquad (19)$$

It turns out that a reasonable qualitative understanding of the widths and masses of ρ and ω can simultaneously be achieved through such a procedure, though the mass of ω comes out rather low on this model.

Finally, we would like to report briefly on a model of SU_3 symmetric interaction within the separable interaction framework.[19] SU_3 interactions[20] have become very fashionable in the classification of the various resonances being daily discovered. A lot of work on this subject has been done by various workers, starting with Cutkosky.[21] We will consider here a very simple type of interaction, viz., a vertex formed by three-vector mesons. The approximate mass degeneracy of the vector meson octet $(\rho \omega K^* \bar{K}^*)$, as well as the fact that there is only a unique form of V^3-coupling satisfying SU_3 (the so-called F-type coupling) makes the vector meson octet a convenient basis set in terms of which it is possible to study the spin-parity structures of V-V composites. We take the V^3-vertex of the form

$$H(x) = iG \sum_{\alpha\beta\gamma} f_{\alpha\beta\gamma} [\boldsymbol{\nabla} \times \mathbf{V}_\alpha(x)] \cdot [\mathbf{V}_\beta(x) \times \mathbf{V}_\gamma(x)] \qquad (20)$$

where $f_{\alpha\beta\gamma}$ are the structure constants of the octet representation of SU_3.[20] A vertex like (20) generates Möller interactions between V-V pairs in the standard way. The tensor part of the V-V interaction has the spin-dependence

$$S_{12}(\mathbf{K}) = (\mathbf{S}_1 \cdot \mathbf{K})(\mathbf{S}_2 \cdot \mathbf{K}) - (\mathbf{S}_1 \cdot \mathbf{S}_2)K^2$$
$$\mathbf{K} = \mathbf{p} - \mathbf{q} = \text{momentum transfer} \qquad (21)$$

The isotopic factor $[G_T]$ for all the possible channels forms a purely numerical matrix depending entirely on the group structure in (T, Y) space. Our model of the interaction is now specified by taking

$$\langle \mathbf{p} | H | \mathbf{q} \rangle = 2\lambda \{G_T\} S_{12}(\mathbf{K}) v(p) v(q) \qquad (22)$$

where the radial factors are assumed separable. Using this model through a Schrödinger equation of the form

$$(p^2 - k^2)\Psi_T(\mathbf{p}) = -\int d^3\mathbf{q} \langle \mathbf{p} | H | \mathbf{q} \rangle \Psi_T(\mathbf{q}) \qquad (23)$$

we have examined various spin-parity states of $T = 0$ and find that only the states 0^{++}, 0^{-+}, 1^{--}, 1^{+-}, 2^{++} are attractive. The observed ϕ and f^0 seem to fit into this scheme semiquantitatively with octet and 27-fold representations, respectively. However, the mass of the state 0^{-+} comes appreciably higher than observed. As for the 1^{+-} state of $T = 0$, it has an exact counterpart (in our model) in the $T = 1$ state as well, as it is interesting that some evidence for a $T = 1$ axial vector meson decaying into $\pi + \omega$ has been recently reported.[22]

In conclusion I would say that a dynamic scheme through a Schrödinger-type equation and separable kernels seems to have at least a small domain of validity within which it provides a good deal of advantage in calculation and does not require many approximations.

REFERENCES

1. Y. Yamaguchi, *Phys. Rev.* **95**: 1635 (1954).
2. A. N. Mitra and V.L. Narasimhan, *Nuclear Phys.* **14**: 407 (1960).
3. A.N. Mitra and J.H. Naqvi, *Nuclear Phys.* **25**: 307 (1961).
4. J.H. Naqvi, *Nuclear Phys.* **36**: 578 (1963).
5. A.N. Mitra and S.P. Pandya, *Nuclear Phys.* **20**: 455 (1960).
6. V.L. Narasimhan *et al.*, *Nuclear Phys.* **23**: 529 (1962).
7. A.N. Mitra, *Phys. Rev.* **123**: 1892 (1961).
8. A.N. Mitra and J.D. Anand, *Phys. Rev.* **130**: 2117 (1963).
9. A.N. Mitra, *Nuclear Phys.* **32**: 529 (1962).
10. R. Blankenbecler, *Phys. Rev.* **122**: 983 (1961).
11. A.N. Mitra, *Phys. Rev.* **131**: 832 (1963).
12. A.N. Mitra and V.S. Bhasin, *Phys. Rev.* **131**: 1265 (1963).
13. A.N. Mitra and Shubha Ray, *Ann. Phys.* **21**: 439 (1963).
14. B.S. Bhakar, *Nuclear Phys.* **46**: 572 (1963).
15. Y. Yamaguchi and Y. Yamaguchi, *Phys. Rev.* **95**: 1639 (1954).
16. B.S. Bhakar and A.N. Mitra (to be published).
17. F.J. Dyson *et al.*, *Phys. Rev.* **95**: 1644 (1954).
18. A.N. Mitra, *Phys. Rev.* **127**: 1342 (1962).
19. A.N. Mitra and S.R. Choudhury, *Phys. Letters* **7** (No. 1) (1963).
20. M. Gell-Mann, *Phys. Rev.* **125**: 1067 (1962); Y. Néeman, *Nuclear Phys.* **26**: 222 (1961).
21. R.E. Cutkosky *et al.*, *Phys. Letters* **1**: 93 (1963).
22. Abolins *et al.*, *Phys. Rev. Letters* **2**: 381 (1963).

Form Factors of the Three-Nucleon Systems ^{3}H and ^{3}He

T. K. RADHA*

MATSCIENCE
Madras, India

Report prepared by G. RAMACHANDRAN

Recently, Schiff[1] has discussed the theory of ^{3}H and ^{3}He form factors in connection with experiments being done at Stanford on elastic scattering of high-energy electrons from these nuclei.[2] Both ^{3}H and ^{3}He have bound-state spin $= \frac{1}{2}$, and an analysis of the electron-scattering data on these nuclei has been made using the Rosenbluth formula, from which one obtains four form factors, namely, charge and moment form factors for ^{3}H and the same for ^{3}He. The experimental results showed that the two moment form factors and the charge form factor for ^{3}H are quite similar to each other, while the charge form factor for ^{3}He falls off somewhat more rapidly. These observations fit in with a simple intuitive explanation in terms of the spatial distributions of the like pairs of nucleons. Since the spins of the like pair (protons in ^{3}He and neutrons in ^{3}H) are primarily opposite, the moment is carried mainly by the odd nucleon in both cases. Also, the charge in the case of ^{3}H is carried by the odd nucleon, while in the case of ^{3}He the charge is carried by the like particles. Thus, the experimental observations suggest a distribution for each of the like pair which is different from that of the odd nucleon. This could be pictured as arising from the different binding of the odd nucleon, which could be bound to each of the like

---

* Now on leave at the Department of Physics, Stanford University, Palo Alto, California.

ones by a linear combination of the triplet and singlet interactions that is more strongly attractive than the singlet interaction binding the like pair, thereby accounting for a more extended distribution in space of the like pair than that of the odd nucleon. For example, the ground state could be a $^2S_{\frac{1}{2}}$ state symmetric in the space coordinates of the like pair but not symmetric with respect to one member of the like pair and the odd nucleon. This is, however, inconsistent with charge independence of nuclear forces. Blatt and Derrick[3] have analyzed all the possible ground states of the three-nucleon systems, according to which there are three possible $^2S_{\frac{1}{2}}$ states:

1. A state (S), fully symmetric in the space coordinates of all three nucleons, which is the dominant state (96 % from binding energy calculations).
2. A state that is antisymmetric in the space coordinates of any pair of the three nucleons.
3. A state (S') of mixed symmetry.

Schiff has shown that the seeming inconsistency with the principle of charge independence discussed earlier corresponds to the interference terms between the $^2S_{\frac{1}{2}}$ states (S and S') of different symmetry in an exact isotopic spin formulation.

Among the possible states of the three-nucleon systems enumerated by Sachs,[4] and Blatt and Derrick, there are, in addition to the $^2S_{\frac{1}{2}}$ states, three $^2P_{\frac{1}{2}}$ states, a $^4P_{\frac{1}{2}}$ state, and three $^4D_{\frac{1}{2}}$ states which together could occur with greater probability than the nonsymmetric $^2S_{\frac{1}{2}}$ state (S'), since an appreciable mixture of the D-states could be expected on account of the tensor interactions. The P-states and the completely antisymmetric $^2S_{\frac{1}{2}}$ state are not expected to be present in the ground-state wave function to any appreciable extent. One could therefore expect the SD cross-term to be a dominant correction. The SD cross-terms do not contribute to the charge form factor because of orthogonality between the spin-$\frac{1}{2}$ and spin-$\frac{3}{2}$ states, but contribute to the moment form factor. Here again, only the spin part of the magnetic moment operator contributes, which vanishes for zero momentum transfer because of the orthogonality between the S- and D-states. On the other hand, the observed magnetic moments (i.e., at zero momentum transfer) of these nuclei do not fit with such a description and one has either to incorporate a large percentage of the P-states (which is incompatible with binding energy calculations) or assume the existence of an exchange-current contribution to the magnetic moment.

Denoting by F_X the SD cross-term contribution plus the exchange-current contribution to the moment form factor, the following relations are obtained (assuming that F_X has the same form for the two nuclei):

$$F_{^3\text{H}}^E = 2F_L F_N^E + F_0 F_P^E$$

$$2F_{^3\text{He}}^E = 2F_L F_P^E + F_0 F_N^E$$

$$\mu_{^3\text{H}} F_{^3\text{H}}^M = (\mu_{^3\text{H}} - \mu_P) F_X + \mu_P F_0 F_P^M + \tfrac{2}{3}\mu_N F_N^M (F_0 - F_L)$$

$$\mu_{^3\text{He}} F_{^3\text{He}}^M = (\mu_{^3\text{He}} - \mu_N) F_X + \mu_N F_0 F_N^M + \tfrac{2}{3}\mu_P F_N^M (F_0 - F_L)$$

where F_L and F_0 are body form factors arising from the S-state and SS' cross-term (to be precise, they are the linear combinations $F_L = F_1 - \tfrac{1}{2}F_2$, $F_0 = F_1 + \tfrac{2}{3}F_2$, F_2 being the cross-term and $F_2 \ll F_1$) $F_N^{E,M}$ and $F_P^{E,M}$ are the nucleon electromagnetic form factors, and the μ denote the magnetic moments. The F_X contribution could be determined empirically. The work that is being done concerns the determination of the S'-state probability, which can be determined from the rate of capture of slow neutrons by deuterium. At thermal energies, the transition occurs through the magnetic dipole transition and the rate of capture depends essentially on the S'-state contribution. The wave function for the continuum $n + d$ is taken to fit the n-d scattering data well. The best fit comes from the Gaussian wave function. Taking, therefore, the Gaussian function for the final triton, the transition rate is calculated. The transition rate proceeding through the exchange-moment term (where the main contribution comes from the S-state) is also being estimated. Calculations are not yet finalized.

REFERENCES

1. L.I. Schiff, (To be published.)
2. H. Collard, R. Hofstadter, A. Johansson, R. Parks, M. Ryneveld, A. Walker, M.R. Yearian, R.B. Day, and R.T. Wagner, *Phys. Rev. Letters* **11**: 132 (1963).
3. G. Derrick and J.M. Blatt, *Nuclear Phys.* **8**: 310 (1958).
4. R.G. Sachs, *Nuclear Theory*, Addison-Wesley Publishing Co., Inc., Cambridge, Massachusetts, 1953.

Muon Capture by Complex Nuclei

V. DEVANATHAN

UNIVERSITY OF MADRAS
Madras, India

It has been well established[1] that the muon possesses characteristic Bohr orbits of its own around the nucleus and that the trapping of the muon into these orbits via ordinary atomic interaction is the precursor of any specific reaction with the nucleus. It is also known from the early experiments of Conversi, Pancini, and Piccioni[2] that in light elements a negative μ-meson in the K-orbit generally decays before it is captured by the nucleus. Only for $Z > 10$ does capture become more probable than decay. Thus, the experiments become not only difficult but also unreliable for low $Z(Z < 10)$, because the free μ-decay rate is then greater than the μ-capture rate, and, in fact, one often measures the sum of the two rates. So the heavier nuclei are favored for the study of muon capture by the experimentalists, but the theoretical study is rendered difficult by the complexity of the structure of the heavier nuclei. The interaction Hamiltonian of the basic elementary processes $\mu^- + p \rightarrow n + \nu$ is not yet well known. Although the universal Fermi interaction is assumed to be operative in the weak processes such as β-decay, μ-decay, and μ-capture, only the μ-decay is free from the effects of strong interactions; μ-capture is affected by the presence of strong interaction currents[3] which considerably alter the structure of the interaction Hamiltonian. The fundamental universal "bare" four-fermion interaction, which is assumed to be a V-A interaction, is modified by the effects of strong interaction to yield other types of interactions, namely, pseudoscalar, scalar, tensor, and weak magnetism. The determination of these coupling constants is of primary interest. For this, one requires extensive experimental data such as (1) partial transition rates in muon capture (toward definite states of the final nucleus), (2) angular distribution of neutrons (after capture of polarized muons), (3) polarization of neutrons emitted, and (4) radiative muon capture.

The interpretation of experimental results with complex nuclei depends on (a) the precise magnitude and form of the muon-capture interaction and (b) the details of nuclear structure. One can, however, select experiments with nuclei that will yield the maximum possible information concerning the moun capture interaction with a minimum of detailed knowledge about their structure, e.g., nuclei with closed shells for either the proton or neutron or both. Double magic nuclei such as ^{16}O and ^{40}Ca are most convenient for this purpose. Alternatively, muon capture may be used as a tool to obtain the information concerning nuclear structure.

1. STRUCTURE OF THE S-MATRIX ELEMENT

We suppose that μ-capture can be described by the following interaction Lagrangian with axial vector and vector couplings:

$$
\begin{aligned}
\mathscr{L}_I = {} & z_2 f_A \bar{\psi}_\nu (1 - \gamma_5) i\gamma_\lambda \gamma_5 \psi_\mu \bar{\psi}_n i\gamma_\lambda \gamma_5 \psi_p \\
& + z_2 f_V \bar{\psi}_\nu (1 - \gamma_5) \gamma_\lambda \psi_\mu \bar{\psi}_n \gamma_\lambda \psi_p + \text{Hermitian conjugate}
\end{aligned}
\tag{1}
$$

where f_A and f_V are the unrenormalized coupling constants, z_2 is the nucleon wave-function renormalization constant (a corresponding constant for the lepton is set equal to unity, since we will treat the weak interaction to the lowest order and also neglect the electromagnetic effects).

To the lowest order in weak interactions, the S-matrix element is given by

$$
S = i(2\pi)^4 \delta(n + p_\nu - p - p_\mu)M
\tag{2}
$$

where n, p_ν, p, and p_μ are, respectively, the four-momenta of the neutron, neutrino, proton, and muon. The matrix element M includes the effect of strong interactions and it can be written as

$$
\begin{aligned}
M = {} & \bar{u}_\nu (1 - \gamma_5) i\gamma_\lambda \gamma_5 u_\mu \langle n \mid P_\lambda \mid p \rangle \\
& + \bar{u}_\nu (1 - \gamma_5) \gamma_\lambda u_\mu \langle n \mid V_\lambda \mid p \rangle
\end{aligned}
\tag{3}
$$

where $\mid n \rangle$ and $\mid p \rangle$ are physical neutron and proton states and

$$
\begin{aligned}
P_\lambda &= z_2 f_A \bar{\psi}_n i\gamma_\lambda \gamma_5 \psi_p \\
V_\lambda &= z_2 f_V \bar{\psi}_n \gamma_\lambda \psi_p
\end{aligned}
\tag{4}
$$

The general structure of the matrix elements of P_λ and V_λ may be deduced, following Berman. The matrix element must be a four-vector and must therefore be a linear combination of

$$p_\lambda, \; n_\lambda, \; \gamma_\lambda, \; \sigma_{\lambda\rho}p_\rho, \text{ and } \sigma_{\lambda\rho}n_\rho$$

or, equivalently,

$$q_\lambda = p_\lambda - n_\lambda, \; P_\lambda = p_\lambda + n_\lambda, \; \gamma_\lambda, \; \sigma_{\lambda\rho}P_\rho \text{ and } \sigma_{\lambda\rho}q_\rho$$

where

$$\sigma_{\lambda\rho} = \tfrac{1}{2}(\gamma_\lambda\gamma_\rho - \gamma_\rho\gamma_\lambda)$$

The matrix element $\langle n \,|\, V_\lambda \,|\, p \rangle$ can be written as

$$\begin{aligned}
\langle n \,|\, V_\lambda \,|\, p \rangle = \bar{u}(n)\{&f_1(q^2)q_\lambda + f_2(q^2)P_\lambda + f_3(q^2)\gamma_\lambda \\
&+ f_4(q^2)\sigma_{\lambda\rho}P_\rho + f_5(q^2)\sigma_{\lambda\rho}q_\rho\}u(p)
\end{aligned} \tag{5}$$

It can now be shown that of the five terms, only three are independent:

$$\begin{aligned}
\bar{u}(n)\sigma_{\lambda\rho}q_\rho u(p) &= \tfrac{1}{2}\bar{u}(n)\{(\gamma_\lambda\gamma_\rho - \gamma_\rho\gamma_\lambda)(p_\rho - n_\rho)\}u(p) \\
&= \tfrac{1}{2}\bar{u}(n)\{\gamma_\lambda \slashed{p} - \gamma_\lambda \slashed{n} - \slashed{p}\gamma_\lambda + \slashed{n}\gamma_\lambda\}u(p) \\
&= \tfrac{1}{2}\bar{u}(n)\{2M\gamma_\lambda - \gamma_\lambda \slashed{n} - \slashed{p}\gamma_\lambda\}u(p) \\
&= \tfrac{1}{2}\bar{u}(n)\{4M\gamma_\lambda - 2P_\lambda\}u(p)
\end{aligned} \tag{6}$$

using the relations

$$(\slashed{p} - M)u(p) = 0$$
$$\bar{u}(n)(\slashed{n} - M) = 0$$

and

$$\slashed{p}\gamma_\lambda = -\gamma_\lambda \slashed{p} + 2p_\lambda$$

Thus, the P_λ term may be included in the γ_λ and $\gamma_{\lambda\rho}q_\rho$ terms. Similarly, we have

$$\bar{u}(n)\sigma_{\lambda\rho}P_\rho u(p) = -2\bar{u}(n)q_\lambda u(p) \tag{7}$$

and so the $\sigma_{\lambda\rho}P_\rho$ term can be absorbed into the term, leaving us with only three independent terms. We thus obtain the general structure

$$\begin{aligned}
\langle n \,|\, P_\lambda \,|\, p \rangle = \bar{u}(n)\{&Ai\gamma_\lambda\gamma_5 - Bq_\lambda\gamma_5 \\
&+ E\sigma_{\lambda\rho}q_\rho\gamma_5\}u(p)
\end{aligned} \tag{8}$$

$$\langle n \,|\, V_\lambda \,|\, P \rangle = \bar{u}(n)\{C\gamma_\lambda - iD\sigma_{\lambda\rho}q_\rho + iFq_\lambda\}u(p)$$

Hence, the complete matrix element (which includes the effects of the strong interactions of the nucleons) for the process of muon capture can be written as

$$\begin{aligned}
M = \frac{1}{\sqrt{2}}[\bar{u}_\nu(1 - \gamma_5)&i\gamma_\lambda\gamma_5 u_\mu)\{A(\bar{u}_n i\gamma_\lambda\gamma_5 u_p) \\
&- B(\bar{u}_n q_\lambda\gamma_5 u_p) + E(\bar{u}_n\sigma_{\lambda\rho}q_\rho\gamma_5 u_p)\} \\
&+ (\bar{u}_\nu(1 - \gamma_5)\gamma_\lambda u_\mu)\{C(\bar{u}_n\gamma_\lambda u_p) \\
&- iD(\bar{u}_n\sigma_{\lambda\rho}q_\rho u_p) + iF(\bar{u}_n q_\lambda u_p)\}]
\end{aligned} \tag{9}$$

The quantities A, B, C, D, E, and F are form factors which depend on the invariant four-momentum transfer q^2, and they are nearly constants if q^2 does not vary too much. Besides, they are real if T-invariance holds.

Weinberg[4] distinguishes between interactions of the first class (the terms with coefficients A,B,C,D) and those of the second class (the terms with coefficients E, F). Goldberger and Treiman[3] and Fujii and Primakoff[5] assume explicitly that there are no second-class interactions; however, this should be established experimentally.

A, B, C, D, E, and F may be considered as a kind of apparent coupling constants. The following notation is introduced such that all coupling constants have the usual dimension of the four-fermion coupling constant:

$C = g_V$ vector apparent coupling constant.
$A = g_A$ axial vector apparent coupling constant.
$m_\mu B = g_P$ (induced) pseudoscalar coupling constant.
$2MD = g_M$ weak magnetism coupling constant.
$m_\mu F = g_S$ (induced) scalar coupling constant.
$2ME = g_T$ (induced) tensor coupling constant.

The mass of the muon is m_μ and M is the nucleon mass.

It is assumed that the fundamental universal "bare" four-fermion interaction is a V-A interaction with coupling constants $g_V^0 = -g_A^0$ (having the same value for beta-decay, muon decay, and muon capture). The strong interactions of the nucleons cause renormalizations of the nucleon currents such that additional terms appear and also alter the values of A and C. (In the absence of strong interaction, $A = g_A^0$, $C = g_V^0$, $B = D = E = F = 0$.)

2. THE EFFECTIVE HAMILTONIAN FOR MUON CAPTURE

Using the four-momentum conservation law $p_\lambda - n_\lambda = \nu_\lambda - \mu_\lambda$ and the free Dirac equation for u_ν, u_μ, u_p, and u_n, namely,

$$(\gamma_\lambda p_\lambda - im)u = 0$$

and also the relation

$$p_\lambda p_\lambda = -m^2$$

we obtain

$$M = \frac{1}{\sqrt{2}}\{ A[\bar{u}_\nu(1 - \gamma_5)i\gamma_\lambda\gamma_5 u_\mu](\bar{u}_n i\gamma_\lambda\gamma_5 u_p)$$

$$+ m_\mu B[\bar{u}_\nu(1 - \gamma_5)\gamma_5 u_\mu](\bar{u}_n\gamma_5 u_p)$$

$$- \frac{iC}{M}[\bar{u}_\nu(1 - \gamma_5)\gamma_\lambda u_\mu](\bar{u}_n p_\lambda u_p)$$

$$- \frac{iC}{2M}[\bar{u}_\nu(1 - \gamma_5)\gamma_\lambda(\mu_\lambda - \nu_\lambda)u_\mu](\bar{u}_n u_p) \qquad (10)$$

$$+ i\left(\frac{C}{2M} + D\right)[\bar{u}_\nu(1 - \gamma_5)\gamma_\lambda(\mu_\rho - \nu_\rho)u_\mu]$$

$$\times (\bar{u}_n\sigma_{\lambda\rho}u_p)\}$$

In the above, we have neglected the terms involving E and F.

The transition matrix element M calculated in a nonrelativistic approximation for the muons and nucleons corresponds to a suitably chosen nonrelativistic effective Hamiltonian $\mathscr{H}_{\text{eff}}$. The $\mathscr{H}_{\text{eff}}$, generalized to a many-nucleon problem, is related to the corresponding matrix element by

$$M = \left\langle \begin{array}{c} \text{final state} \\ \text{of} \\ A \text{ nucleons} \end{array} \right| \mathscr{H}_{\text{eff}} \left| \begin{array}{c} \text{initial state} \\ \text{of} \\ A \text{ nucleons} \end{array} \right\rangle \qquad (11)$$

and describes muon capture by an aggregate of A dressed nucleons. We find for $\mathscr{H}_{\text{eff}}$, in a configuration space representation,

$$\mathscr{H}_{\text{eff}} = \frac{1}{\sqrt{2}}\bar{u}_\nu\tau^+\frac{1 - \vec{\sigma}\cdot\hat{\nu}}{\sqrt{2}}\sum_{i=1}^{A}\tau_i^{(-)}[G_\nu l\cdot l_i$$

$$+ G_A\vec{\sigma}\cdot\vec{\sigma}_i - G_P(\vec{\sigma}\cdot\hat{\nu})(\vec{\sigma}_i\cdot\hat{\nu}) \qquad (12)$$

$$- \frac{g_V}{MC}(\vec{\sigma}\cdot\hat{\nu})(\vec{\sigma}\cdot\vec{p}_i) - \frac{g_A}{MC}(\vec{\sigma}\cdot\hat{\nu})(\vec{\sigma}_i\cdot\vec{p}_i)]u_\mu$$

This may be written in a more compact form as follows:

$$\mathscr{H}_{\text{eff}} = \tfrac{1}{2}\bar{u}_\nu\tau^+(1 - \vec{\sigma}\cdot\hat{\nu})(A + \vec{\sigma}\cdot\vec{B})u_\mu \qquad (13)$$

where

$$A = \sum_{i=1}^{A}\left\{ G_V l\cdot l_i - \frac{g_V}{MC}\hat{\nu}\cdot\vec{p}_i\right\}$$

$$\vec{B} = \sum_{i=1}^{A}\left\{ G_A\vec{\sigma}_i - G_P\hat{\nu}(\vec{\sigma}_i\cdot\hat{\nu})\right. \qquad (14)$$

$$\left. -i\frac{g_V}{MC}(\hat{\nu}\times\vec{p}_i) - \frac{g_A}{MC}\vec{\nu}(\vec{\sigma}_i\cdot\vec{p}_i)\right\}$$

In the above, $\tau_i^{(-)}$, l_i, $\vec{\sigma}_i$, and $\vec{p}_i$ are operators for the nucleon; τ^+, $\vec{\sigma}$, and l are the lepton operators; $\hat{\nu}$ is the unit vector in the direction of the neutrino momentum $\vec{\nu}$; and $\tau^{(-)}$ and $\tau^{(+)}$ are the operators decreasing and increasing the charge of the nucleon or lepton by one unit. G_ν, G_A, and G_P might be called "effective coupling constants," and they are given by

$$G_V = C\left(1 + \frac{\nu}{2M}\right) = g_V^{(\mu)}\left(1 + \frac{\nu}{2M}\right)$$

$$G_A = A - \left(\frac{C}{2M} + D\right)\nu$$

$$= g_A^{(\mu)} - (g_V^{(\mu)} + g_M^{(\mu)})\frac{\nu}{2M} \tag{15}$$

$$G_p = (m_\mu B - A)\frac{\nu}{2M} - \left(\frac{C}{2M} + D\right)\nu$$

$$= [(g_P^{(\mu)} - g_A^{(\mu)}) - (g_V^{(\mu)} + g_M^{(\mu)})]\frac{\nu}{2M}$$

In equation (12), only $1/M$ terms are included, and terms of higher orders are neglected.

It will be instructive to inquire as to the extent to which the μ-capture coupling constants differ from the β-decay coupling constants. If we assume the universal Fermi interaction, then the coupling constants which are functions of four-momentum transfers will differ in these two cases due to the difference in the four-momentum transfer. Arguing thus, Fujii and Primakoff have obtained the following relations:

$$g_V^{\mu C} = 0.97\, g_V^\beta \qquad g_A^{\mu C} = g_A^\beta \tag{16}$$

The conserved vector current hypothesis yields the value of the weak magnetic coupling constant:

$$g_M = \frac{\mu_p - \mu_n}{g_V} \approx 3.7 g_V. \tag{17}$$

where μ_p and μ_n are the proton and neutron anomalous magnetic moments in nuclear magnetons.

Theoretical considerations of Wolfenstein and of Goldberger and Treiman provide a plausible value for the induced pseudoscalar coupling constant:

$$g_p \approx 8 g_A \tag{18}$$

In neglecting the terms involving E and F, we have explicitly assumed

$$g_S = 0 \qquad g_T = 0 \tag{19}$$

Relations (16) to (19) are to be tested experimentally, and they may provide a test of the hypothesis of the universal Fermi interaction and that of the conserved vector current.

3. MUON CAPTURE RATES

The capture rate is given by

$$\Gamma = \frac{\nu^2}{2\pi} \varphi_\mu^2(0) M^2$$

where (20)

$$\varphi_\mu^2(0) = \frac{(Zm_\mu e^2)^3}{\pi}$$

for a point nucleus and m_μ is the meson mass and Ze the charge of nucleus. Averaged over initial muon states and summed over final neutrino states,

$$M^2 = AA^* + BB^* \tag{21}$$

A sum over final and average over initial nuclear states is implicit here. Retaining only the $1/M$ terms, we obtain

$$AA^* = |G_V|^2 \left| \int l \right|^2 - \frac{G_V g_V}{MC} \left[\left(\int l \right) \left(\hat{\nu} \cdot \int \vec{p} \right) + C \cdot C \right]$$

$$BB^* = |G_A|^2 \left| \int \vec{\sigma} \right|^2 + (G_P^2 - 2G_P G_A) \left| \hat{\nu} \cdot \int \vec{\sigma} \right|^2$$

$$- \frac{G_A g_V}{MC} i \hat{\nu} \cdot \left[\left(\int \vec{\sigma} \right) \times \left(\int \vec{p} \right)^* + C \cdot C \right]$$

$$- \frac{G_A g_A - G_P g_A}{MC} \left[\left(\hat{\nu} \cdot \int \vec{\sigma} \right)^* \left(\int \vec{p} \cdot \vec{\sigma} \right) + C \cdot C \right]$$

where

$$\int l = \langle f | \sum_i \tau_i^{(-)} \exp\left(-i\vec{\nu} \cdot \vec{r}_i\right) | i \rangle$$

$$\int \vec{\sigma} = \langle f | \sum_i \tau_i^{(-)} \exp\left(-i\vec{\nu} \cdot \vec{r}_i\right) \vec{\sigma}_i | i \rangle$$

$$\int \vec{p} = \langle f | \sum_i \tau_i^{(-)} \exp\left(-i\vec{\nu} \cdot \vec{r}_i\right) \vec{p}_i | i \rangle \tag{23}$$

$$\int \vec{p} \cdot \vec{\sigma} = \langle f | \sum_i \tau_i^{(-)} \exp\left(-i\vec{\nu} \cdot \vec{r}_i\right) \vec{p}_i \cdot \vec{\sigma}_i | i \rangle$$

If we omit the velocity-dependent terms (which can be neglected in the first instance), then we get for the partial capture from the given initial to the specified final state

$$\Lambda_{\mu C}(i \rightarrow f) = \frac{\nu^2}{2\pi} \int \frac{d\hat{\nu}}{4\pi} \left\{ G_V^2 \left| \int l \right|^2 + G_A^2 \left| \int \vec{\sigma} \right|^2 \right.$$

$$\left. + (G_P^2 - 2G_P G_A) \left| \hat{\nu} \cdot \int \vec{\sigma} \right|^2 \right\} \tag{24}$$

For nuclei with closed proton and closed neutron shells, we have the following relation:

$$\int \frac{d\hat{v}}{4\pi} \left| \hat{v} \cdot \int \vec{\sigma} \right|^2 = \frac{1}{3} \left| \int \vec{\sigma} \right|^2 \qquad (25)$$

Using relation (25), we obtain

$$\Lambda_{\mu C}(i \rightarrow f) = \frac{v^2}{2\pi} \int \frac{d\hat{v}}{4\pi} \left[G_F^2 \left| \int 1 \right|^2 + G_{GT}^2 \left| \int \vec{\sigma} \right|^2 \right] \qquad (26)$$

with

$$G_F = G_V$$
$$G_{GT} = G_A^2 + \tfrac{1}{3}(G_P^2 - 2G_P G_A) \qquad (27)$$

Since

$$\sum_f \int \frac{d\hat{v}}{4\pi} \left| \int \vec{\sigma} \right|^2 = 3 \sum_f \frac{d\hat{v}}{4\pi} \left| \int 1 \right|^2 \qquad (28)$$

the total capture rate for doubly closed shell nuclei, e.g., ^{16}O and ^{40}Ca, becomes

$$\Lambda_{\mu C} = \frac{v^2}{2\pi} |\varphi_\mu|^2 [G_F^2 + 3G_{GT}^2]M^2 \qquad (29)$$

where

$$M^2 = \sum_f \int \frac{d\hat{v}}{4\pi} |\langle f| \sum_i \tau_i^{(-)} \exp(-i\hat{v} \cdot \vec{r}_i)|i\rangle|^2 \qquad (30)$$

Thus, from the total capture rate of the doubly closed shell nuclei, we can determine the combination of the coupling constants $G_F^2 + 3G_{GT}^2$. For the separate determination of the Fermi and Gamow–Teller coupling constants, the partial rates have to be observed.

The matrix elements (30) and (23) can be evaluated[6-8] using either the statistical or shell model of the nucleus. Primakoff, Tolhock, and Duck have evaluated these matrix elements, and the details of the calculation are given in the references cited. If we require the total capture rate, we can with advantage use the closure approximation. It is generally expected that the shell model calculations will yield more reliable results, and in this case one can obtain the total capture rate by enumerating all the possible final states and summing over the partial rates obtained for each specified final state of the nucleus.

Recently, Cohen et al.[9] have experimentally observed the separate μ-capture transitions between discrete states of ^{16}O and ^{16}N with a view to obtaining information about the basic interaction involved. The partial capture rates to the four relevant states of ^{16}N are given on the top of the next page:

Relevant State of ^{16}N	Capture Rate (sec^{-1})
1^-	$(1.73 \pm 0.10) \times 10^3$
0^-	$(0.66 \pm 0.11) \times 10^3$
2^-	$(6.76 \pm 0.71) \times 10^3$
3^-	unobserved
Total	$(9.15 \pm 0.71) \times 10^3$

The partial μ-capture rate for the transition to 3^- is negligible. The rate for the transition from the 0^+ ground state of ^{16}O to a 0^- state of ^{16}N is particularly sensitive to the assumed magnitude of the induced pseudoscalar constant (g_P). The experimental value, Cohen et al.[9] point out, seems to suggest a value

$$g_P \approx 15 g_A$$

as against the value $g_P \approx 8 g_A$ suggested by Goldberger and Treiman.

REFERENCES

1. Wheeler, *Rev. Mod. Phys.* **21**: 133 (1949).
2. M. Conversi, E. Pancini, and O. Piccioni, *Phys. Rev.* **71**: 209 (1947).
3. M.L. Goldberger and S.B. Treiman, *Phys. Rev.* **111**: 354 (1958).
4. S. Weinberg, *Phys. Rev.* **112**: 1375 (1959).
5. A. Fujii and H. Primakoff, *Nuovo Cimento* **12**: 327 (1959).
6. H. Primakoff, *Rev. Mod. Phys.* **31**: 802 (1959).
7. J.R. Luyton, H.P.C. Rood, and H.A. Tolhock, *Nuclear Phys.* **41**: 236 (1963).
8. I. Duck, *Nuclear Phys.* **35**: 27 (1962).
9. R.C. Cohen, S. Devans, and A.D. Kaman, *Phys. Rev. Letters* **11**: 134 (1963).

Electrodynamics of Superconductors

NEW YORK UNIVERSITY
New York, New York

---

1. INTRODUCTION

We shall begin with the three basic facts concerning the electrodynamics of superconductors: (A) *Persistent currents*—Flow of currents in certain metals in the superconducting phase seemed to indicate that these metals behave like perfect conductors. (B) *The Meissner effect*—However, the expulsion of magnetic field from the bulk of superconductor clearly indicated that a superconductor is a good diamagnet. (C) *Flux quantization*—If a multiply connected superconductor, say, a ring, is in a magnetic field and later cooled below the transition temperature, the expelled flux trapped inside the hole does not have arbitrary values but assumes values which are multiples of a unit quantity. This phenomenon throws great light on the microscopic theory of superconductivity. Flux quantization has been anticipated by London and Onsager.

Among the macroscopic phenomenological theories, the most successful ones are London's theory and the Landau–Ginzburg theory; a brief description of them follows.

London's Theory

London assumes that the current in a superconductor is the sum of two terms:

$$J = J_n + J_s \tag{1.1}$$

where J_n and J_s are the normal current and the supercurrent

$$J_n = \sigma E \tag{1.2}$$

and

---

* Department of Physics.

$$J_s = -\frac{1}{c\Lambda} A^{\mathrm{tr}} \tag{1.3}$$

The conductivity of the metal is σ, and Λ is a constant which will be identified below. It is to be noted that (1.3) is true only in a special gauge defined by div $A^{\mathrm{tr}} = 0$. If we use $H = \mathrm{curl}\, A$, we have

$$\mathrm{curl}\, J_s = -\frac{1}{c\Lambda} H \tag{1.4}$$

Equation (1.4) implies the Meissner effect. To see this, we also consider Maxwell's equation

$$\mathrm{curl}\, H = \frac{4\pi}{c} J_s \tag{1.5}$$

In (1.5) we have omitted terms on the right-hand side which are not of interest here. Taking curl on both sides of (1.5) and using (1.4) and div $H = 0$, we have

$$\nabla^2 H = \frac{4\pi}{c\Lambda^2} H \tag{1.6}$$

Consider an infinite half-space (see Fig. 1). With the boundary conditions indicated, there will be two solutions for (1.6), one exponentially increasing and the other exponentially decreasing. Only the latter has physical meaning. We have for the field inside the superconductor

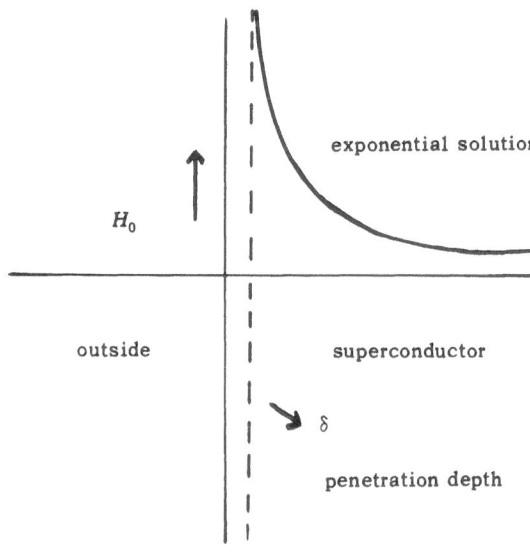

Fig. 1

$$H = H_0 \exp\left(-\frac{x}{\delta}\right) \tag{1.7}$$

The δ is called the *penetration depth*:

$$\delta = \left(\frac{\Lambda e^2}{4\pi}\right)^{1/2} \tag{1.8}$$

Usually, $\delta \sim 10^{-5}$ or 10^{-6} cm. Measurement of δ leads to a determination of Λ.

The relation between the current and the field given by (1.3) is a local one. Pippard suggested a nonlocal relation that is more valid in certain situations. From microscopic considerations, we will be able to derive Pippard's relation, which is of the form

$$J(x) = \int K(x - x')A^{\mathrm{tr}}(x')dx' \tag{1.9}$$

If we assume that $A^{\mathrm{tr}}(x)$ varies very slowly with x, we can recover London's equation (1.3). Writing Pippard's relation in momentum space, we obtain

$$J(q) = K(q)A^{\mathrm{tr}}(q) \tag{1.10}$$

If $K(q)$ has a finite limit as $q \to 0$, we have a superconductor. However, if $K(q) \to 0$ as $q \to 0$, we have a normal metal.

We will now turn to another interesting feature of the superconductor, which is the flux quantization. In any simply connected superconductor the fluxoid given by

$$\oint (A + c\Lambda J)dl = \Phi \tag{1.11}$$

is identically zero. However, in a multiply connected superconductor, this need not be zero, and in fact it is found to be quantized.

Landau–Ginzburg Theory

In this theory, the state of the superconductor is described by means of the wave function ψ, which is also a function of temperature. $|\psi(x, T)|^2$ is taken to represent the density of superphase. We assume for the current, somehow, the usual expression

$$J = -\frac{ieh}{2m}(\psi^* \nabla \psi - \psi \nabla \psi^*) - \frac{e^2}{mc}\psi^* \psi A \tag{1.12}$$

In the presence of field we have for the free-energy density F

$$F = F_n + \alpha(T)|\psi|^2 + \frac{\beta(T)}{2}|\psi|^4 + \frac{h^2}{2m}\left|\left(\nabla - \frac{ie}{hc}A\right)\psi\right|^2 + \frac{H^2}{8\pi} \tag{1.13}$$

We vary ψ to obtain the minimum value for F and thus obtain an equation for ψ:

$$\left\{-\frac{\hbar^2}{2m}\left(\nabla - \frac{ieA}{\hbar c}\right)^2 - \alpha(T) + \beta(T)|\psi|^2\right\}\psi = 0 \qquad (1.14)$$

In the absence of field, that is, when $A = 0$, we have $|\psi|^2 = \alpha(T)/\beta(T)$, which is a constant independent of x. It is easy to see that in the transverse gauge $\nabla \cdot A = 0$, ψ has a rigidity property, that is, ψ does not change for weak perturbations A^{tr} (to first order). With this assumption, we find that the current is given by

$$J = -\frac{e^2}{mc}\psi^*\psi A^{\mathrm{tr}} \qquad (1.15)$$

In obtaining (1.15) we have used the fact that $\nabla \cdot A = 0$ and $(A \cdot \nabla)\psi = 0$ as ψ is constant because of the above assumption. Putting $\psi^*\psi = n_s$, we obtain London's equation:

$$J = -\frac{n_s^2 e^2}{mc}A^{\mathrm{tr}} \qquad (1.16)$$

Hence, we now identify Λ as

$$\Lambda = \frac{m}{e^2 n_s} \qquad (1.17)$$

From the measurement of penetration depths it is possible to estimate n_s, or equally α/β, since δ is

$$\delta = \left(\frac{mc^2\beta}{4\pi e^2 \alpha}\right)^{1/2} \qquad (1.18)$$

Knowing the critical field and the transition temperature, it is easy to calculate α and β separately from the expression

$$H_c^2 = \frac{4\pi\alpha^2}{\beta} \qquad (1.19)$$

which is also valid in this theory.

In this theory, the fluxoid quantization follows immediately, since

$$A + c\Lambda J = -\frac{i\hbar c}{2e}\nabla \log\frac{\psi}{\psi^*} \qquad (1.20)$$

If we put $\psi = |\psi|e^{is}$

$$\oint(A + c\Lambda J)dl = \text{difference in phases at the end of a closed path}$$

$$= \frac{hc}{e}[s] \qquad (1.21)$$

Since the phase has to be a multiple of 2π, the integral should equal $(hc/e)\,n$.

Since ψ in the Landau–Ginzburg theory is an effective wave function, e and m need not refer to mass and charge of a single electron. In fact, we know from experiments that e and m really refer to $2e$ and $2m$. Onsager seems to have been the first to suggest the occurrence of $2e$ in the quantization of flux. Yang has shown generally that in fermion systems only pairs of fermions can condense.

2. GENERALIZED CANONICAL TRANSFORMATION

In a recent paper[1] we have attempted a systematic derivation of these electrodynamic effects using a generalized version of the Bogoliubov transformation. We start with the Hamiltonian

$$\mathcal{H} = \Sigma K_{f_1 f_2} a_{f_1}^* a_{f_2} + \tfrac{1}{2} \Sigma P_{f_1 f_2 f_3 f_4} a_{f_1}^* a_{f_2}^* a_{f_3} a_{f_4} \tag{2.1}$$

where $f = (p, \sigma)$, $2m = h = c = 1$.

In equation (2.1),

$$K_{f_1 f_2} = \langle f_2 | (p - eA)^2 - \mu | f_1 \rangle \tag{2.2}$$

In the Bardeen–Cooper–Schrieffer theory, the attraction between the electrons is the effective balance between the Coulomb repulsion and the effective attractions caused by exchange of phonons. Assuming the metal to be a superconductor, one uses the canonical transformation of Bogoliubov and Valatin and obtains the quasi-particle picture with the new operators α_p:

$$\alpha_p = u_p a_{p\uparrow} + v_p a_{-p\downarrow} \tag{2.3}$$

The new ground state is defined by

$$\alpha_p |0\rangle = 0 \tag{2.4}$$

To obtain the thermodynamics one constructs the density matrix of these "free" quasi-particles and proceeds. The above transformation is sufficient if there is translational invariance in the system. However, when we have a magnetic field, we no longer have translational invariance as well as momentum conservation. Thus we are forced to consider a more generalized transformation of the type

$$\alpha_f = \sum_{f'} \{ u_{ff'} a_{f'} + v_{ff'} a_{f'}^* \} \tag{2.5}$$

In order that this transformation be canonical, we have to impose some conditions on the matrices u and v. The choice of u and v can be adjusted to ensure gauge invariance of the α. The discussion of the gauge invariance of the interaction P is complicated.

To study the theory at finite temperature, we introduce the density matrix defined by

$$W = \prod_f (\tfrac{1}{2} + \Gamma_f)\alpha_f^* \alpha_f + (\tfrac{1}{2} - \Gamma_f)\alpha_f \alpha_f^* \qquad (2.6)$$

Now

$$-TS = \frac{1}{\beta}\langle \ln W \rangle \qquad (2.7)$$

where the average of any operator 0 is denoted by $\langle 0 \rangle$ and is given by

$$\langle 0 \rangle = \mathrm{Tr}\, 0W \qquad (2.8)$$

with $\mathrm{Tr}\, W = 1$. The free energy is $\langle H \rangle - TS$ and we minimize with respect to variations of u, v, and Γ.

In doing this, we need

$$G = \langle a^* a \rangle \qquad F = \langle a, a \rangle \qquad (2.9)$$

We also find

$$\langle a_1^* a_2^* a_3 a_4 \rangle = (G + \tfrac{1}{2})_{14}(G + \tfrac{1}{2})_{23} - (G + \tfrac{1}{2})_{13}(G + \tfrac{1}{2})_{24} + F_{12}^+ F_{34} \qquad (2.10)$$

$$\langle \mathcal{H} \rangle = \tfrac{1}{2}\mathrm{Tr}\,\{(E + K)(G + \tfrac{1}{2}) + F^+ D\} \qquad (2.11)$$

$$E = K + Q\cdot(G + \tfrac{1}{2}) \qquad (2.12)$$

$$D = P\cdot F \qquad (2.13)$$

where

$$Q_{f_1 f_2 f_3 f_4} = P_{1342} - P_{1324}$$

We find it convenient to define the following supermatrices:

$$\mathbf{G_2} = \begin{vmatrix} G^* & F \\ -F^* & G \end{vmatrix} \qquad \mathbf{E} = \begin{vmatrix} E^* & D \\ -D^* & E \end{vmatrix} \qquad \mathbf{\Gamma} = \begin{vmatrix} \Gamma & 0 \\ 0 & -\Gamma \end{vmatrix} \qquad (2.14)$$

The Γ variation, i.e., $\delta\,\Gamma$, leads to

$$\Gamma = -\tfrac{1}{2}\tanh\frac{\beta}{2}\epsilon_d \qquad (2.15)$$

where ϵ is only diagonal, and $\epsilon = \mathbf{C E C}^+$ and

$$\mathbf{C} = \begin{vmatrix} u & v \\ v^* & u^* \end{vmatrix} \qquad (2.16)$$

We now note that variations u and v are not arbitrary, since they have to form a canonical transformation. Variation of u and v leads to

$$[\mathbf{G}, \mathbf{E}] = 0 \qquad (2.17)$$

Consider the case when $A = 0$. Now u and v are diagonal. We have

$$E_p = p^2 - \mu + \sum_{p'} (P_{p-p'} - 2P_0)\left(\frac{\Gamma_{p'}}{\epsilon_{p'}} E_{p'} + \tfrac{1}{2}\right) \qquad (2.18)$$

$$D_p = -\sum_{p'} P_{p-p'} \frac{\Gamma_{p'}}{\epsilon_{p'}} D_{p'} \qquad \text{(gap equation)} \qquad (2.19)$$

Assuming that (2.19) possesses a nontrivial solution, we obtain

$$\epsilon_p = (E_p^2 + D_p^2)^{1/2}$$

and

$$\Gamma_p = -\tfrac{1}{2}\tanh\left(\frac{\beta}{2}\,\epsilon_p\right) \qquad (2.20)$$

When $A \neq 0$, we have, to first order in A, after linearization,

$$\left\{ \begin{pmatrix} \rho^- & 0 \\ 0 & \rho^+ \end{pmatrix} + \begin{pmatrix} fpf - gqg & fpg + gqf \\ gpf + fqg & gpg - fqf \end{pmatrix} \right\} \begin{pmatrix} X \\ Y \end{pmatrix}$$

$$= \begin{pmatrix} gK' \\ -fK' \end{pmatrix} \qquad (2.21)$$

where

$$K' = -(2p + q)\cdot eA(q) \qquad (2.22)$$

$$g_p = u_p u_{p+q} + v_p v_{p+q}$$

$$f_p = u_p v_{p+q} - v_p u_{p+q} \qquad (2.23)$$

and with $q = p^* - p$

$$\rho_p^\pm = \frac{\epsilon_p \pm \epsilon_{p+q}}{\Gamma_p \pm \Gamma_{p+q}} = \frac{\epsilon^\pm}{\Gamma^\pm} \qquad (2.24)$$

We can calculate the bulk current to be

$$\langle J'(q)\rangle = \frac{2e}{V}\sum_p (2p + q)(g_p X_p - f_p Y_p) \quad \frac{2e^2 N}{V} A(q) \qquad (2.25)$$

Starting from this expression for current, we can study the various electrodynamic properties of the superconductor. (The details are contained in a previous article.[1]) In this way a fully gauge invariant derivation of the London–Pippard theory is possible. Furthermore, in the approximation in which the potential is slowly varying, it is also possible[2] to derive the Landau–Ginzburg equations.

REFERENCES

1. D. Uhlenbrock and B. Zumino, *Phys. Rev.* **133**: A350 (1964).
2. B. Zumino and D.A. Uhlenbrock, *Nuovo Cimento* **33**: 1446 (1964).

"Temperature Cutoff" in Quantum Field Theory and Mass Renormalization

S. P. MISRA*

REGIONAL ENGINEERING COLLEGE
Rourkela, Orissa, India

1. INTRODUCTION

In quantum field theory, we quite often need a cutoff, particularly if we wish to consider the mass differences of highly symmetric multiplets, in which case we cannot dispose of infinite constants simply by calling them "unobservable effects." Furthermore, interactions of widely diverse strength and structure such as strong, electromagnetic, and weak interactions all require a cutoff of the same order of magnitude, although there is no quantitative relationship between the different effects of a given interaction. Therefore, the cutoff for which we are searching is not likely to be of mathematical origin, if field theory is in fact a good approximation, giving us at least quantities of the right order of magnitude.

Since our search for a cutoff at high momenta can lead us outside the realm of present assumptions, we consider the possibility of finding one based on some physical principle. There is such a cutoff in statistical mechanics, universal for an assembly of particles at fixed temperature and independent of the nature of the individual particles. Quantum mechanics may be interpreted probabilistically or statistically, and so perhaps we can acquire a "temperature cutoff" by analogy.

This paper is an examination of the above analogy to determine whether it is sensible or irrelevant.

All processes of quantum mechanics involve the matrix multiplication rule, where we put

---

* Department of Mathematics.

$$I = \sum \prod_j |n'_j \vec{p}'_j\rangle\langle n'_j \vec{p}'_j|$$

with n'_j giving the eigenvalues of the number operator for momentum $\vec{p}'_j$. We replace this operator, up to one-particle states, by

$$\rho = |O\rangle\langle O| + \sum_j F(\vec{k}_j)a^*(\vec{k}_j)|O\rangle\langle O|a(\vec{k}_j)$$

where $F(\vec{k}_j)$ is the statistical weight of the state $a^*(\vec{k}_j)|O\rangle$. We assume that the weight is given by a Maxwellian distribution and make the concept fully relativistic by defining a 4-vector β^μ corresponding to $\beta = (kT)^{-1}$ in an arbitrary frame of reference and taking $F(k) = \exp(\beta^\mu k_\mu + m)$, with the energy–momentum 4-vector always on the mass shell. For scattering problems, it is difficult to make use of equilibrium, but drawing an analogy from thermodynamics and considering fluctuations can give amusing and useful results.

In solving the problems, we consider electromagnetic effects only and deviate somewhat from the standard perturbation technique by making use of field equations, the completeness of free-particle states, and translational invariance. The intermediate particles are always taken as physical particles, since translational invariance is applicable only to the physical state, and since in a few models[1,2] we find that real particle states and bare particle states are orthogonal. Further, our intermediate states lie on the mass shell; there is covariance, but no conservation of either energy or momentum. This approach to the second order which we have considered is like the standard approach in content, although different in concept, and it suits our purpose.

With this approach, we have achieved reasonable agreement for the electron mass being completely due to electromagnetic interactions, and for the π^+–π^0 mass difference, the neutron–proton mass difference, and electron–nucleon scattering, using the same "temperature cutoff" for all these processes. We also note that the statistical interpretation of quantum mechanics is a consequence of using statistical assembly for the intermediate states in the transitions from initial to final states.

2. ELECTRON AND PION MASS RENORMALIZATION

In order to anticipate isotropy of direction in space, we consider a one-electron state having zero momentum. Eliminating space–time dependence by translational invariance, we obtain by perturbation analysis from the electron field equation (notations are from Jauch and Rohrlich[3])

$$(-m' + m)(2\pi)^{-3/2}u_r(\vec{o})$$

$$= ie\gamma^\mu \int [\langle O | \psi(O)a_s^*(\vec{k}) | O \rangle F_e(\epsilon_k) \langle a_s(\vec{k})A_\mu(O)a_r^*(\vec{o}) | O \rangle$$

$$+ \langle O | \psi(O) a_r^*(\vec{o})a_\lambda^+(\vec{k}) | O \rangle F_p(k) \langle O | a^\lambda(\vec{k})A_\mu(O) | O \rangle] d^3k \quad (1)$$

where m' is the renormalized mass and $F_e(\epsilon_k) = \exp[-\beta(\epsilon_k - m')]$ and $F_p(k) = \exp(-\beta k)$ are the weights of intermediate-state electrons and photons, with $k = |\vec{k}|$. Summation over repeated indices is understood.

The photon field equation then gives us, after some elementary calculations,

$$\langle O | a_s(\vec{k})A_\mu(O)a_r^*(\vec{o}) | O \rangle = -ie(2\pi)^{-3}\sqrt{\frac{m'}{\epsilon_k}} \frac{u_s(\vec{k})\gamma_\mu u_r(\vec{o})}{2m'(\epsilon_k - m')}$$

such that

$$-ie\gamma^\mu \langle O | \psi(O)a_s^*(\vec{k}) | O \rangle \langle O | a_s(\vec{k})A_\mu(O)a_r^*(\vec{o}) | O \rangle$$

$$= (2\pi)^{-3/2} \frac{e^2}{(2\pi)^3} \frac{u_s(\epsilon_k - 2m')u_r(\vec{o})}{2m'\epsilon_k(\epsilon_k - m')} \quad (2)$$

In equation (2) we have omitted an odd term in $\vec{k}$ in the numerator, which will not contribute in the subsequent symmetric integration.

Similarly, also simplifying the second term on the right-hand side of equation (1), we finally get

$$m' - m = \frac{e^2}{(2\pi)^3}\int d^3k \left[\frac{(k + m' - 2m)[F(p)(k)]}{k(2m'k + m'^2 - m^2)} \right.$$

$$\left. - \frac{(\epsilon_k - 2m')F_e(\epsilon_k)}{2m'\epsilon_k(\epsilon_k - m')} \right] \quad (3)$$

To see how equation (3) differs from the usual perturbation expression, we note that when we take the same cutoff on the right-hand side of this equation we obtain a linear divergence multiplied by $m' - m$ as the cutoff goes to infinity.[4] This term cannot occur when we substitute $m' = m$, i.e., expand in terms of either the physical mass or the bare mass as in the usual theory, which would give only the remaining logarithmic divergence in equation (3). For a large and finite cutoff, this approximation does not seem reasonable, since the term with $m' - m$ is multiplied by the large cutoff factor.

We next assume that $\beta^{-1} \gg m'$ and m. Then equation (3) gives us, retaining the leading terms,

$$m' - m = \frac{\alpha}{\pi m'}\int_0^\infty dk \left[\left(k + \frac{m'^2 - 4mm' + m^2}{2m'} \right) e^{-\beta k} \right.$$

$$\left. - (k - m')e^{-\beta(k-m')} \right]$$

where $\alpha = e^2/4\pi$. When the bare mass is taken as zero, putting $m_e = m'$, we get $m_e = (\alpha/2\pi)\beta^{-1}$, such that $\beta^{-1} = 860\, m_e = 440$ MeV.

The mass of the electron could arise from such a small cutoff because of the linear divergence term, which is always omitted. However, the faith we have put on second-order perturbation calculations may not be justified. Also, we have taken for granted the possibility that the mass arises in spite of the invariance of the original Lagrangian by assuming that all the solutions need not maintain the symmetries of the Lagrangian.[5, 6]

For subsequent analyses, we shall instead take

$$\beta^{-1} = 410\,\text{MeV} \tag{4}$$

which does not introduce much error in the preceding calculations and which gives good agreement with the π^+–π^0 mass difference, which is more reliable.

Similar calculations give this mass difference with cutoff (4) as

$$\mu' - \mu = \frac{2\alpha\mu'^2}{\mu' + \mu}\left[(\beta\mu')^{-2} + (\beta\mu')^- + \tfrac{1}{2}I\right] = 9.7 m_e$$

where

$$I = \frac{1}{\mu'^2}\int_0^\infty \frac{(\omega_k - \mu')^2 \exp\left[-\beta(\omega_k - \mu')\right]}{\omega_k}\, dk = 7.02$$

3. RELATIONSHIP WITH FEYNMAN DIAGRAMS AND n-p MASS DIFFERENCE

The above procedure is relativistically covariant, since we have assumed the electron and the pion to be in the rest frame (where complete symmetry can be expected; see also the following section), and we can go over to any frame of reference by a Lorentz transformation. However, the method may also be related to Feynman diagrams for comparison and convenience of calculations.

For this purpose, we note that the electron mass renormalization term in Feynman theory is

$$m' - m = \frac{ie^2}{(2\pi)^4}\int \frac{\gamma_\mu[i\gamma(p - k) - m]\gamma^\mu}{[(p - k)^2 + m^2]k^2}\, d^4k$$

On first integrating k^0, we write the above contribution as

$$\frac{e^2}{(2\pi)^3}\int d^3k$$

(the sum of the residues in the upper half-plane) of $[2i\gamma(p - k) + 4m]/$
$[(p - k)^2 + m^2]$

$$= \frac{e^2}{(2\pi)^3} \int d^3k \left[\frac{k + m' - 2m}{(2m'k + m'^2 - m^2)k} - \frac{\epsilon_k - 2m'}{2m'\epsilon_k(\epsilon_k - m')} \right] \qquad (5)$$

where the first term contains m' from p, and where we are to substitute $m' = m$, but have written down the two terms according to
a rule to be defined subsequently. The contribution is evaluated in the
rest frame of the electron. In expression (5), the two terms on the
right are the same as those in equation (3), without the corresponding
cutoff factors.

In Feynman theory, we make all the expansions in terms of m'
only. However, in the perturbation method described here, we have to
distinguish the double role of the propagators mixed together in
Feynman diagrams, as may be clear from calculations of equation (3).
One originates from the field equations corresponding to the bare
mass, and the other from the nonlinearity of the field equations in
evaluating some contributions with intermediate particle states. Thus,
in $[i\gamma(p - k) + m]^{-1}(k^2)^{-1}$, when we consider the pole of $[i\gamma(p - k) + m]^{-1}$, (k^2) arises in the denominator from the photon field equation
and not from any intermediate virtual state insofar as use of the field
equations is allowed. Similarly, when we consider the pole of $(k^2)^{-1}$,
$[i\gamma(p - k) + m]^{-1}$ has its origin from the electron field equation. The
approximation we have used corresponds to taking the bare mass in
the propagator when the contribution does not arise from a pole of
the denominator, and taking the physical mass for the contribution
from the poles, when subsequent integration is involved. The terms on
the right-hand side of expression (5) are written on this basis. In both
the terms, of course, p is the 4-momentum of the physical particle.

Another justification for taking the bare mass in the above calculations may be seen as follows: For mass renormalization we must have

$$(\gamma\partial + m)\langle O | \psi(x)a_r^*(\vec{p}) | O \rangle \simeq (-m' + m)\langle O | \psi(x)a_r^*(\vec{p}) | O \rangle$$

But conventional perturbation analysis also takes for the second
term on the right-hand side of equation (1)

$$(\gamma\partial + m')\langle O | \psi(x)a_r^*(\vec{p})a_\lambda^+(\vec{k}) | O \rangle$$
$$= (m' - m)\langle O | \psi(x)a_r^*(\vec{p})a_\lambda^+(\vec{k}) | O \rangle$$
$$- ie \langle O | \psi(x)A_\mu(x)a_r^*(\vec{p})a_\lambda^+(\vec{k}) | O \rangle$$

and rejects the contribution from the first term on the right-hand
side above as negligible or of higher order. In the above analysis, we

have retained this term and found that the corresponding effect in the denominator may be quite significant. Also, if we take the Fourier transform of the above equation, it becomes obviously uncomfortable, for any physical consideration whatsoever, to neglect this term until equation (3), and at this stage the approximation is no longer sensible.

The present analysis thus brings out a difficulty in perturbation method not encountered in power series expansion; as is usual in physics, we can only seek *a posteriori* justification for any truth in such calculations. Actually, our method is no worse in this respect than conventional perturbation analysis, since we merely retain a term that is considered negligible.

The above analogy with Feynman diagrams enables us to calculate the mass renormalization terms easily with the appropriate weight factors introduced at the poles.

Applying this relationship to the Feynman diagrams, we can calculate the neutron–proton mass difference $(57\frac{1}{2})$ and find this to be 2.53 MeV, about double the experimental value, but having the correct sign.

4. ELECTRON–PROTON SCATTERING

In the preceding section, we have considered the cutoff of virtual particles in the rest frame of one real particle, where we imagine vacuum to consist of weakly interacting virtual particles in statistical equilibrium. We shall try to generalize these ideas to include scattering problems by considering the probability of change of particle states given by fluctuations of energy and momentum of the states.

Let us consider the scattering amplitude originating from the interaction term

$$e(\bar{\psi}_P\gamma^\mu\psi_P - \bar{\psi}_e\gamma^\mu\psi_e)A_\mu$$

where ψ_P and ψ_e are the proton and electron field operators, respectively. With p_i, p_f and P_i, P_f as the initial and final 4-momenta of the electron and proton, respectively, we obtain an explicit contribution to the matrix element

$$\langle p_f P_f |\bar{\psi}_P\gamma^\mu\psi_P A_\mu| p_i P_i\rangle$$

as

$$\sim \langle p_f P_f |\bar{\psi}_P| p_f\rangle\gamma^\mu\langle p_f |\psi_p| P_i p_f\rangle$$
$$\times \square^{-1}[\langle p_f P_i |\bar{\psi}_e| P_i\rangle\gamma^\mu\langle P_i |\psi_e| p_i P_i\rangle] \qquad (6)$$

where the last part is written from the electron current part of the photon field equation.

Next, for electron mass renormalization, β may be replaced by the time-like 4-vector $\beta^\mu = (\beta p^\mu)/m'$ in an arbitrary frame of reference, and $\exp\left[-\beta(\epsilon p' - m')\right]$ of the rest frame will be replaced by $\exp(\beta^\mu \Delta p_\mu) = \exp\{\beta[(pp'/m') + m']\}$. β^μ has zero space component when the total momentum vanishes. To obtain a general result with this, let us take n_j to represent the covariant probability of the state $|\vec{p}_j\rangle$. Then, in an arbitrary frame of reference, we consider

$$(\textstyle\sum n_j \epsilon_j)^2 - (\textstyle\sum n_j \vec{p}_j)^2 = \text{constant}$$

which for a variation of n_j gives us

$$E \sum \epsilon_j \delta n_j - \vec{P} \sum \vec{p}_j \delta n_j = 0$$

where E and $\vec{P}$ are the total energy and momenta of the system. The above equation together with the equation

$$\textstyle\sum \ln n_j \delta n_j = \sum \delta n_j = 0$$

gives us

$$n_j = A \exp(BP^\mu p_{j\mu}) \qquad (7)$$

The above result is trivial, but it clearly demonstrates that we may regard $(KT)^{-1} = \beta$ as being replaced by $\beta^\mu = BP^\mu$ and the covariant product taken in equation (7).

Thus, for the case of electron-proton scattering, it seems reasonable to write

$$\beta^\mu = \lambda(p^\mu + P^\mu)$$

where $\lambda = \lambda(s)$ may depend on energy, with $s = -(p + P)^2$. If we take λ as a constant, we obtain

$$\lambda = \frac{\beta}{m + M} \simeq \frac{\beta}{M}$$

since, if the particles were at rest, we must have $\beta^\mu = (\beta, 0, 0, 0)$.

Let us now consider the matrix element (6), where we have a fluctuation of the energy–momentum of the electron and then also of the proton. Let $\ln \Omega(p, P)$ denote the entropy of the state $|p, P\rangle$ and $\omega(p, P)$ the intrinsic probability of the state. Then we take

$$\omega(p, P) = \Omega(p, P) \exp\left[\beta^\mu(p_\mu + p_\mu)\right]$$

except for a multiplicative constant, where, introducing thermodynamic analogy with entropy, we shall evaluate[8]

$$\ln \frac{\omega(p + \Delta p, P)}{\omega(pP)} = \ln \Omega(p + \Delta p, P) - \ln \Omega(p, P) + \beta^\mu \Delta P_\nu$$

using the covariant entropy equation

$$d \ln \Omega(p, P) = -\beta^\mu(dp_\mu + dP_\mu) \tag{8}$$

Equation (8) gives us

$$\ln \Omega(p + \Delta p, P) - \ln \Omega(p, P)$$

$$= \frac{\partial \ln \Omega}{\partial p_\mu} \Delta p_\mu + \tfrac{1}{2} \frac{\partial^2 \ln \Omega}{\partial p_\mu \partial p_\nu} \Delta p_\mu \Delta p_\nu$$

$$= -\beta^\mu \Delta p_\mu - \tfrac{1}{2} \frac{\partial \beta^\mu}{\partial p_\nu} \Delta p_\mu \Delta p_\nu$$

Hence, retaining up to second power in $t = -(p_f - p_i)^2 = -\Delta p^\mu \Delta p_\mu$, we obtain the probability factor for the above energy-momentum fluctuation as

$$\exp\left(\tfrac{1}{2} \lambda t - \tfrac{1}{4} \frac{d\lambda}{ds} t^2\right)$$

which depends only on the momentum transfer of the particle when λ is a constant. Taking the same further contribution for the fluctuation of the proton state, we obtain the extra factor in the matrix element (6) as $\exp(\lambda t)$. For the Hamiltonian including magnetic moment terms, the same intermediate states will arise. Hence, if λ is a constant we get the form factor

$$F(t) = \exp\left(\beta \frac{t}{M}\right) \tag{9}$$

which gives the mean square radius as $6\beta/M$, such that, by equation (4)

$$\langle r^2 \rangle^{1/2} = 0.78 \times 10^{-13} \, \text{cm} \tag{10}$$

This result agrees well with experimental results from low-energy electron-nucleon scattering, with identical values for all the form factors. It is anticipated that deviations from the above Gaussian distribution arise because of pionic effects, which are necessarily small.

Equation (9) has some further interesting consequences. For high-energy (inelastic) nucleon–nucleon scattering, we notice that the fluctuation of the initial 4-momentum P_1 of the nucleon to a final 4-momentum P_1' will have a factor $\beta^\mu = (\beta/2M)(P_1^\mu + P_2^\mu)$

$$F(t) = \exp\left[-\frac{\beta}{4M}(P_1 - P_1')^2\right] \tag{11}$$

But

$$-(P_1 - P_1')^2 = 2M^2 - 2E_1 E_1' + 2|\vec{P}_1||\vec{P}_1'|\cos\theta$$
$$= 2M^2 - 2E_1(E_1' - |\vec{P}_1'|) - 2|\vec{P}_1'|(E_1 - |\vec{P}_1|) - 2|\vec{P}_1||P_1'|(1 - \cos\theta)$$

$$(12)$$

From equation (12), it is obvious that if $E_1 \gg M$, E_1'/E_1 must remain finite and nonzero as $E_1 \to \infty$, and, further, that $\cos\theta$ must approach unity fast enough to make factor (11) a nonvanishing contribution. This means that for high-energy collisions, where P_2 is at rest, the incoming particle will lose a finite fraction of its energy and will proceed without any change of direction. The existence of the elasticity parameter thus becomes quite plausible.

Further, in the CM frame of reference, the "temperature" is $\beta E_{cm}/(2M)$, so that the particles produced must be at rest with respect to this frame of reference. This leads to the belief that only "resonances" are produced *at rest* in the CM frame of reference at high energy, and these must decay into mesons having only moderate energies. Also, since a change in quantum number of the incoming particles would statistically imply the creation of fresh particles, this high-energy particle, besides retaining a proportional fraction of energy, will also retain its quantum numbers. Of course, all these conclusions are based on factor (11) without any reference to the dynamics and our concept may be oversimplified.

Besides the amusing agreements listed above, we prefer the model with λ as a constant, since here the crossing relation for the Mandelstam representation will still be valid for two channels (not all three), in a limited class of Feynman diagrams, which may dominate some processes, explaining the success of dispersion relations.

5. DISCUSSION

We feel that the value of dealing with scattering problems by use of any form of the present cutoff hypothesis must depend on the extent to which this method is sensible for mass renormalization itself, and also on the clarification of the theoretical difficulties that arise. For example, we have left completely open the question of γ_5 invariance for electron mass renormalization and do not know how the higher-order terms contribute. It would also be interesting to know whether the cutoff gives rise to any parity-violating weak interactions.[9] For the present, however, we have concentrated on finding out whether there is reasonable experimental evidence for such a hypothesis and have considered only those effects where we may have some reliance on the method of calculation up to high energies.

Besides the close agreement with experimental results, the following broad features regarding mass renormalization and scattering processes should be mentioned. We see that the mass renormalization term breaks up into the difference of two terms, which may be recognized as one of the forward-scattering amplitudes for two different physical processes. Thus, for neutron-proton mass difference (where the necessity of taking bare and physical mass separately does not arise), the mass renormalization term breaks up into the difference of Compton scattering for nucleons and electromagnetic scattering of two identical nucleons. When the above hypothesis is not true, but Feynman diagrams describe the truth in some way (e.g., leading to, dispersion relations), the above observation leads to a clear separation of applying experimental results for scattering to mass renormalization. We can use the size of the nucleons determined from Hofstadter experiments only in the term of mass renormalization which corresponds to electromagnetic scattering of two nucleons; the form factors for Compton scattering should be used in the other term.[10] This means, in conventional language, that we are to introduce separately the form factors for time-like and space-like momentum transfers at the nucleon vertex, when these form factors can be separately known.

REFERENCES

1. L. Van Hove, *Physica* **18**: 145 (1952).
2. S.P. Misra, *Proc. Nat. Inst. Sci. India* **28A**: 261 (1962).
3. J.M. Jauch and F. Rohrlich, *Theory of Photons and Electrons*, Addison-Wesley, 1955.
4. S.P. Misra, "Symposium on Cosmic Rays and Elementary Particles," Madras, India, December 1961.
5. Y. Nambu and G. Jona-Lasino, *Phys. Rev.* **122**: 345 (1961).
6. J. Schwinger, *Phys. Rev.* **128**: 2425 (1962).
7. R.P. Feynman and G. Speisman, *Phys. Rev.* **94**: 500L (1954).
8. J.E. Mayer and M.G. Mayer, *Statistical Mechanics* John Wiley & Sons, 1946, p. 227.
9. L. Van Hove, *Physica* **25**: 365 (1959).
10. R.K. Bose, Unpublished.

Recent Developments in the Statistical Mechanics of Plasmas

HUGH DEWITT

LAWRENCE RADIATION LABORATORY
Livermore, California

1. INTRODUCTION

A plasma is an ionized gas at any density and temperature: for example, stellar interiors. Such a gas differs from every other gas in that each particle feels the effects of all other particles because of the infinite range of the Coulomb force acting between every pair of charged particles. By contrast, in an ordinary nonideal gas, say, argon, each atom feels only the force from its nearest neighbor.

This paper is a brief review of some recent work in the application of modern methods of the quantum mechanical many-body problem to the statistical mechanics of plasmas. The approach is entirely *microscopic* and is most easily described with field-theory methods, in contrast to the macroscopic approach of magnetohydrodynamics, a continuum theory. Generally, the mathematical problems that arise as a result of the Coulomb interaction, i.e., the numerous divergences, can now be quite satisfactorily handled, at least for a nonrelativistic plasma. In a relativistic plasma, of course, quantum electrodynamics must be combined with statistical mechanics, and renormalization must be used to handle some of the divergences.

Some of the problems that may be considered are:

1. Equation of state, i.e., $P/\rho = f(T, \rho)$, where P is the pressure, ρ is the particle number density, and T is the absolute temperature. Laboratory plasmas are usually at such low density that the pressure very nearly obeys the ideal gas law. In stellar interiors, the deviations from the ideal gas law may be quite large.

2. Electrons in metals. Here, one has an essentially zero-tempera-ture plasma, since $kT \ll \epsilon_F$. Using statistical mechanics, one can calculate such things as the correlation energy and the specific heat.

3. Plasma oscillations, i.e., charge density fluctuations which propagate through the plasma.

4. Bremsstrahlung and inverse bremsstrahlung.

5. Relaxation toward thermal equilibrium.

6. Transport coefficients.

At the present time, there are two general methods of attacking these problems, quantum-mechanical perturbation theory and Green's function techniques. In this discussion, I will describe some aspects of the perturbation approach. Plasmas at equilibrium are described by the grand partition function, from which one calculates the grand potential:

$$\beta\Omega(\alpha, \beta) \equiv \beta PV = \log \text{Tr } e^{\alpha N - \beta(H_0 + H_I)} \qquad (1)$$

where

$\beta = 1/kT$, T = absolute temperature.

$\alpha = \beta\mu$, μ = chemical potential.

H_0 = the sum of kinetic energies of the plasma particles.

H_I = the sum of interaction energies among the particles, primarily the sum of pairwise Coulomb interactions.

A basic unifying idea helps in the treatment of all of these problems, namely, the *effective* interaction between any two particles in the plasma. Let us consider an electrically neutral two-component plasma of point electrons and point ions. Charge neutrality requires

$$Z_e N_e + Z_i N_i = 0 \qquad (2)$$

The effective interaction may be looked at as a sum of two parts (Fig. 1):

$$u_S(r_{12}, t_1 - t_2) = \frac{Z_1 Z_2 e^2}{r_{12}} \delta(t_1 - t_2) + u_P(r_{12}, t_1 - t_2) \qquad (3)$$

where the first part is the direct Coulomb interaction and the second is the polarization interaction which contains retardation effects. Fourier analysis of u_S gives

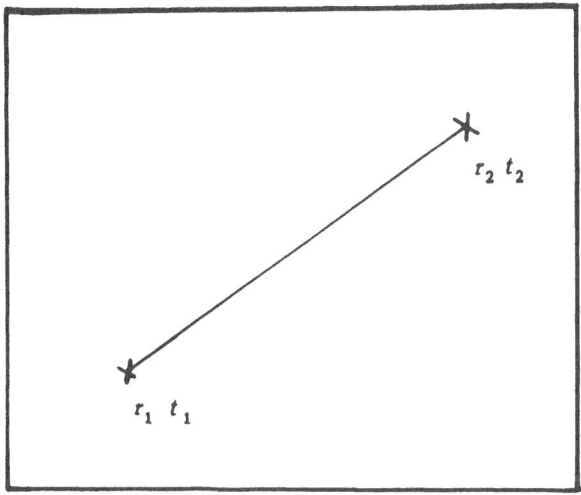

Fig. 1

$$\int\int d^3r\,dt\,e^{i(k\cdot r-\omega t)}u_S(r,t) = u_S(k,\omega)$$
$$= u_C(k) + u_P(k,\omega) \tag{4}$$

with $u_C = (4\pi e^2)/k^2$ for the Coulomb potential. The Fourier transform of u_S may be described diagrammatically in the following manner:

$$u_S(k,\omega) = \frac{}{k} + \mathbf{p}\;\bigcirc\;\mathbf{p+k} + \;\;\cdots \tag{5}$$

$$= \int = \underline{} + \bigcirc$$

$$= \frac{u_C(k)}{\epsilon(k,\omega)}$$

In the first line, u_s is the sum of first the direct Coulomb interaction (horizontal line), interaction via one intermediate plasma particle excited from state $\mathbf{p}$ to state $\mathbf{p} + \mathbf{k}$, interaction through two particles, etc. The bubbles represent the intermediate particles. The sum is represented by a slanting wiggly line, since it can end at different times. The quantity $\epsilon(k, \omega)$ is the longitudinal dielectric function of the plasma, and provides in principle an exact description of how the Coulomb interaction is modified by the plasma intervening between two charges. We may write $\epsilon(k, \omega)$ as

$$\epsilon(k, \omega) = 1 + u_c(k)\lambda(k, \omega) \tag{6}$$

where $\lambda(k, \omega)$ represents the Fourier analysis of a charge-density fluctuation (the bubble in the diagrams). The polarization potential may then be written as

$$u_p(k, \omega) = -u_c(k)u_s(k, \omega)\lambda(k, \omega) \tag{7}$$

It should be pointed out that $\epsilon(k, \omega)$ describes the response of the plasma to an external field. Thus, suppose a fast charged particle is shot into an equilibrium plasma. If the electric field around this particle in a vacuum is $\mathbf{E}_0(\mathbf{r}', t')$, then the electric field set up in the plasma at a space-time point $(\mathbf{r}, t)$ is

$$\mathbf{E}(\mathbf{r}, t) = \int_{-\infty}^{t} dt' \int d^3\mathbf{r}' \epsilon(\mathbf{r} - \mathbf{r}', t - t')\mathbf{E}_0(\mathbf{r}', t') \tag{8}$$

for which the Fourier analysis is

$$\mathbf{E}(k, \omega) = \epsilon(k, \omega)\mathbf{E}_0(k, \omega) \tag{9}$$

Written in this way, it is clear that $\epsilon(k, \omega)$ is the response function of the plasma. A proper treatment of such an explicitly time-dependent problem is best handled with Green's function techniques. At this point, we simply mention that the perturbation expansion of the grand partition function when properly rearranged yields immediately an expression for $u_s(k, \omega)$ and hence $\epsilon(k, \omega)$.

The quantity $\lambda(k, \omega)$ may itself be described diagrammatically for each term of a perturbation expansion as follows:

$$\lambda(k, \omega) = \mathbf{p} \quad \mathbf{p} + \mathbf{k} = \quad + \quad + \quad +$$

$$+ \quad + \quad + \cdots \tag{10}$$

$$= \lambda^{(0)}(k, \omega) + \lambda^{(1)}(k, \omega) + \cdots \lambda^{(n)}(k, \omega) + \cdots$$

The superscript n refers to the number of internal effective interactions, either self-energy effects or exchange interactions. Clearly, for large n it becomes impossible to obtain exact mathematical expressions for $\lambda^{(n)}$. It has been done for $n = 0$ and 1.[1] The first term, $n = 0$, corresponds exactly to the well-known random phase approximation (RPA) introduced by Bohm and Pines in the early 1950's. The expression is

$$\lambda^{(0)} = \sum_{\mathbf{p}} \frac{f^-(p) - f^-(\mathbf{p} + \mathbf{k})}{[(\mathbf{p} + \mathbf{k})^2/2m - p^2/2m] - (\hbar\omega \pm i\epsilon)} \tag{11}$$

where

$$f^-(p) = \frac{1}{\exp[-\alpha + (\beta p^2/2m)] \pm 1} \qquad \begin{array}{l} + \text{ for Fermi–Dirac} \\ - \text{ for Bose–Einstein} \end{array}$$

Using the Dirac rule

$$\lim_{\epsilon \to 0} \frac{1}{X \pm i\epsilon} = p\frac{1}{X} \pm i\pi\delta(X) \tag{12}$$

one may calculate the real and imaginary parts

$$\lambda^{(0)} = \lambda_R^{(0)} \pm i\lambda_I^{(0)} = N\beta[\mathscr{L}_R^{(0)}(z, u) \pm i\mathscr{L}_I^{(0)}(z, u)] \tag{13}$$

and obtain explicit functions for $\lambda_R^{(0)}$ and $\lambda_I^{(0)}$. These look like the graphs shown in Fig. 2, where

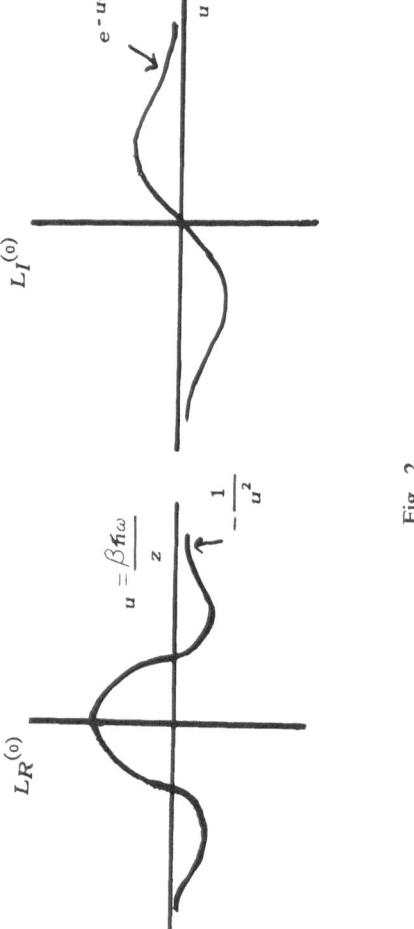

Fig. 2

$$z = \frac{k}{(2mkT)^{1/2}} \quad \text{and} \quad u = \frac{\beta \hbar \omega}{z}$$

In the static limit, $\omega = 0$, the effective interaction becomes

$$u_S(k, 0) = \frac{4\pi e^2}{k^2 + 4\pi e^2 \beta (Z_e^2 \rho_e + Z_i^2 \rho_i)} \tag{14}$$

which in configuration space becomes

$$u_S(r) = \frac{e^2}{r} e^{-r/\lambda_D} = \text{static screened Coulomb potential} \tag{15}$$

where

$$\lambda_D = [4\pi e^2 \beta (Z_e^2 \rho_e + Z_i^2 \rho_i)]^{1/2} = \text{Debye screening length}$$

In most plasma physics work, the static screened potential is used for calculations. The more fundamental and correct form of the effective interaction, $u_S(k, \omega) = u_c(k)/\epsilon(k, \omega)$, allows the introduction of retardation effects which can be significant in some problems.

In general, it is not possible to write $u_S(r, t_2 - t_1)$, so statistical mechanics calculations are best done in momentum space with the use of $u_S(k, \omega)$ in some approximation. Note from the graph of $\lambda_R^{(0)}(k, \omega)$ that for some values of ω the real part of the dielectric function will vanish:

$$\epsilon_R[k, \omega_P(k)] = 0 \tag{16}$$

This equation defines the plasma oscillation frequency $\omega_P(k)$ for which a charge-density fluctuation will propagate through the plasma. This dispersion relation is easily solved for $\omega_P(k)$ to give

$$\omega_P^2(k) = \omega_c^2 + \langle v^2 \rangle k^2 + \cdots \tag{17}$$

with

$$\omega_c = \left(\frac{4\pi e^2 \rho}{m}\right)^{1/2} = \text{classical plasma frequency}$$

$$\langle v^2 \rangle = \frac{3kT}{m} \quad \text{for a nondegenerate plasma}$$

The imaginary part at $\omega_P(k)$, i.e., $\epsilon_I[k, \omega_P(k)]$, then gives the time required for the plasma oscillation to be damped out.

2. EQUATION OF STATE

The grand potential of a plasma may be written as

$$\beta\Omega = \log \text{Tr} \exp \left[\alpha N - \beta\left(\sum_{i=1}^{N} \frac{p_i^2}{2m_i} + \sum_{i \neq j} \frac{Z_i Z_j e^2}{r_{ij}} \right) \right]$$

$$= \beta\Omega_0 + \int_0^{e^2} \frac{de'^2}{e'^2} \langle H_I(e'^2) \rangle \tag{18}$$

$$= \langle N \rangle \left\{ S_0 + S_{\text{ring}} + \sum_{n=2}^{\infty} S_n + \sum_{n=1}^{\infty} S_{n,\text{exchange}} \right\}$$

where $\beta\Omega_0 = \langle N \rangle S_0$ is the quantum ideal gas pressure. The third line in the above equation represents *all* orders of perturbation theory suitably rearranged so as to remove all divergences.[2]

The significant terms for a finite-temperature plasma and not too high density are

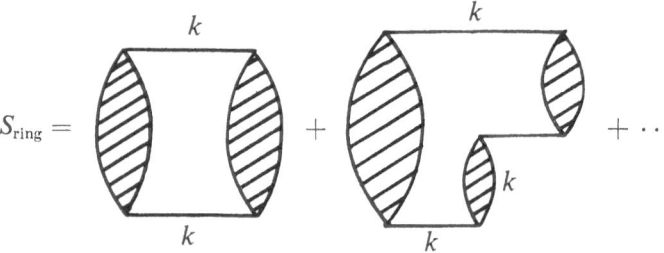

$$\tag{19}$$

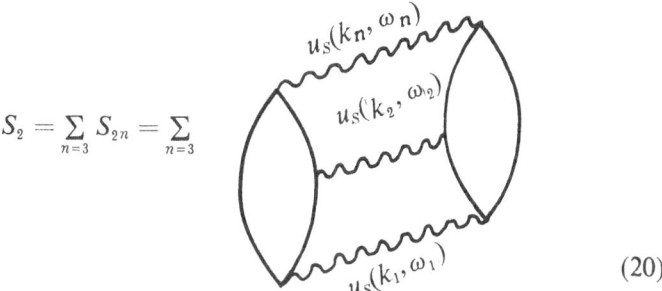

$$S_2 = \sum_{n=3} S_{2n} = \sum_{n=3} \qquad\qquad\qquad (20)$$

In general, the quantity S_n is the contribution to the thermodynamic potential from a cluster of n charges interacting via the general effective interaction any number of times, but subject to the restriction that at least three screened interactions end on every particle of the cluster. The ring diagrams are equivalent to the RPA, while S_2 is the next significant contribution beyond the RPA. Clearly, S_2 represents the sum of ladder diagrams, but with screened interactions $u_S(k, \omega)$ replacing the direct Coulomb interaction for the rungs of the ladders.

Before discussing the evaluation of these terms, it is useful to introduce the appropriate dimensionless parameters for a high-temperature and low-density plasma. Consider first the fundamental lengths of the gas:

$$l_c = \frac{e^2}{\langle KE \rangle} \rightarrow \frac{e^2}{kT} \qquad \text{at high } T$$

$$\lambda = \frac{\hbar}{(2m\langle KE \rangle)^{1/2}} \rightarrow \frac{\hbar}{(2mkT)^{1/2}} \qquad (21)$$

$$\lambda_S = \left(\frac{kT}{4\pi e^2 \rho} \right)^{1/2}$$

At high temperature, these lengths are ordered as

$$l_c < \lambda < \lambda_S$$

Because the thermal de Broglie wavelength λ is greater than the distance of closest approach l_c, it is to be expected that some quantum effects will remain in the plasma thermodynamic functions even as $T \rightarrow \infty$. Thus, for finite $\hbar$, the high-temperature limit is not a classical limit. The dimensionless parameters are the ratios of these lengths:

$$\Lambda = \frac{l_c}{\lambda_S} \propto e^3 \beta^{3/2} \rho^{1/2}$$

$$\gamma = \frac{\lambda}{\lambda_S} \propto e\hbar\beta\rho^{1/2} m^{-1/2} \qquad (22)$$

At high temperature and low density, these parameters are ordered as

$$\Lambda > \gamma > 1$$

High temperature means $kT > Ryd$, in order to satisfy $\Lambda < \gamma$.

From the final diagrammatic form of S_{ring}, one recognizes that only the polarization potential $u_P(k, \omega)$ is involved. With hindsight, one could almost guess how to write the ring contribution to $\beta\Omega$. The actual result was first obtained by Montroll and Ward in 1958.[3] A convenient form to write it is

$$\langle N \rangle S_{\text{ring}} = \frac{1}{2} \int_0^1 \frac{de'^2}{e'^2} \sum_k \sum_{\gamma=-\infty}^{\infty} u_P(k, \beta\hbar\omega = 2\pi i\gamma)\lambda(k, 2\pi i\gamma) \tag{23}$$

$$\gamma = 0, \pm 1, \pm 2, \ldots$$

This form, in which the integral over ω appears as a sum over integers, i.e., pure imaginary frequencies at $\omega_\gamma = 2\pi i\gamma/\beta\hbar$, is a mathematical convenience. The function $\lambda(k, \omega)$ is real for $\omega = \omega_\gamma$, and complex otherwise. Actual evaluation of the integrals and sums in this expression gives [4]

$$S_{\text{ring}} = \frac{\Lambda}{3}P(\gamma) \qquad \text{valid for } \Lambda < \gamma \tag{24}$$

where

$$P(\gamma) = 1 + \sum_{n=1}^{\infty} a_n\gamma^{2n} - \sum_{n=1}^{\infty} b_n\gamma^{2n-1} \qquad \gamma^2 < \gamma_c^2 = 2.042$$

$$= 1 - \frac{\pi^{1/2}}{2^{7/2}}\gamma + \left(\frac{1}{4} + \frac{3}{2s+1}\right)\gamma^2 - \cdots$$

The above result for $P(\gamma)$ gives Ω exact to order e^5. The appearance of $\gamma^2/(2S + 1)$ in $P(\gamma)$ results from the inclusion of self-energy and exchange effects described diagrammatically by

i.e., $\lambda^{(1)}(k, \omega)$

These are also known as local field effects. Note that such terms disappear mathematically in the limit $s \to \infty$, i.e., particles of arbitrarily large spin.

The two-body cluster term S_2 requires a logarithmic cut-off for large r; it appears automatically as λ_s from the form of the effective interaction. Similarly, there is a logarithmic cut-off for small r at either λ or l_c, whichever one is larger. Evaluation of S_2 yields [5]

$$S_2(\Lambda, \gamma) = \frac{\Lambda^2}{6}\left(\log\frac{l_c}{\lambda_S} + D_1\right) \qquad \text{for } \lambda \ll l_c \tag{25}$$

with

$$D_1 = \log 3 + 2C - \frac{11}{6}$$

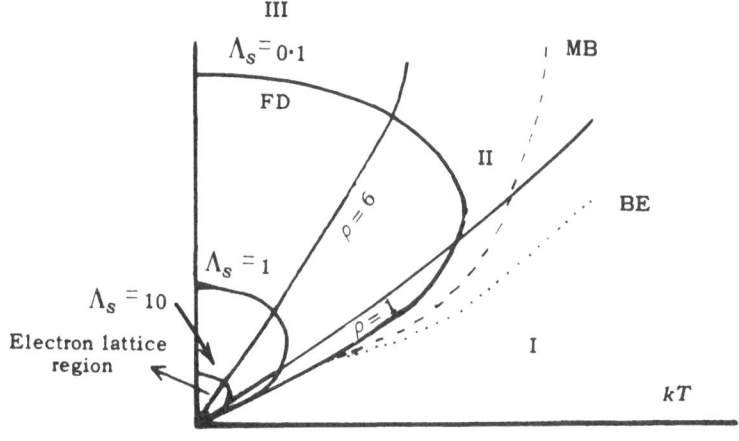

Fig. 3

and

$$S_2(\Lambda, \gamma) \cong S_{23} = \left(\text{◯〰◯} \right) = \frac{\Lambda^2}{6} \left(\log \frac{\lambda}{\lambda_S} + D_2 \right) \qquad (26)$$

$$\text{for } \lambda > l_c \qquad (kT > Ryd)$$

(D_2 is not yet known.)

For a multicomponent gas it is now possible to write the equation of state correct through the logarithm term:

$$\beta PV = (N_e + N_i) \left\{ 1 - \frac{\Lambda}{6} P(\gamma) - \frac{\Lambda^2}{12} \left[Z_e^6 f_e^2 \left(\log \frac{\lambda_{ee}}{\lambda_S} + D_2 \right) \right. \right. $$
$$\left. \left. + 2Z_e^3 Z_i^3 f_e f_i \left(\log \frac{\lambda_{ei}}{\lambda_S} + D_2 \right) + Z_i^6 f_i^2 \left(\log \frac{\beta e^2}{\lambda_S} + D_1 \right) \right] \right\} \qquad (27)$$

In this result the ion–ion term takes the classical form because the ionic mass for real plasmas is at least 2000 times the electron mass, hence

$$\lambda_{ii} \ll \beta e_i^2$$

With quantum statistics required for accurate calculation, it is no longer possible simply to write down the equation of state, although accurate calculations are possible. The ring sum is the dominant contribution to the pressure when a generalized form of the classical interaction parameter is small:

$$\Lambda_S = \frac{1}{4\pi \rho \lambda_S^3} \ll 1 \qquad (28)$$

where

$$\lambda_S^2 = \frac{kT}{4\pi e^2 \rho [1/N(\alpha)][\partial N(\alpha)/\partial \alpha]}$$

$$N(\alpha) = \sum_p \frac{1}{\exp{(-\alpha + \beta\epsilon)} \pm 1}$$

One can mark off regions in the temperature-density plane where sensible calculations can be made. First, draw contours of constant $\zeta = \lambda^3 \rho$ which measure degeneracy ($\zeta \ll 1$ means M–B statistics, while $s \gg 1$ means a very degenerate gas).

For high temperature and low density (Fig. 3), one sees that Λ_S reduces to Λ, while in the high-density low-temperature region, Λ_S becomes $r_S^{3/2}$ where $r_S = 1/\rho^{1/3} a_B$, the conventional parameter used in solid state physics.

Everywhere outside the contour for $\Lambda_S = 0.1$ (Fermi statistics), the ring sum is sufficient to give the interaction corrections to the ideal gas with high accuracy. As one approaches the origin from any direction, the additional terms S_2, S_3, etc., become increasingly important. In the three regions shown above, we have:

Region I (Nondegenerate)

$$\beta PV = \langle N\rangle\{1 - \Lambda P(\gamma)\ldots\} \tag{29}$$

Region II (Partially Degenerate)

Only numerical results from a computing machine are possible because diffraction and exchange effects are inextricably mixed up, so that an analytical evaluation of the ring sum seems impossible.

Region III (Very Degenerate)

The Helmholz free energy $F = E - TS$ may be written as

$$\frac{F}{NRyd} = \left(\underset{\substack{\downarrow \\ \text{ideal gas}}}{\frac{2.21}{r_S^2}} - \underset{\substack{\downarrow \\ \text{exchange}}}{\frac{0.91}{r_S}} + \underset{\substack{\downarrow \\ \text{correlation (Gell-Mann} \\ \text{and Brueckner)}}}{0.93 \log r_S}\right) \tag{30}$$

$$+ \left(\frac{kT}{\epsilon_F}\right)^2 \left(\underset{\substack{\downarrow \\ \text{ideal gas}}}{\frac{\pi^2}{6}} + \underset{\substack{\downarrow \\ \text{(Gell-Mann specific heat)}}}{a r_S \log r_S} + \cdots\right)$$

In the region around the origin enclosed by, say, $\Lambda_S > 10$, the interaction energy is far greater than the kinetic energy and the plasma goes into a lattice. This density–temperature region has some astrophysical applications. For example, in the interior of Jupiter and some protostars, the density and temperature may be such that the protons form a lattice and the electrons form a nearly continuous neutralizing background.

REFERENCES

1. D.F. DuBois, *Ann. Phys.* **7**: 174 (1959). H.E. DeWitt and A.Y. Sakakura (to be published).
2. R. Abe, *Progr. Theoret. Phys.* **22**: 21B (1959).
3. E. Montroll and J. Ward, *Phys. Fluids* **1**: 55 (1958).
4. H.E. DeWitt, *J. Math. Phys.* **3**: 1216 (1962).
5. H.E. DeWitt (to be published.)

Introduction to Quantum Statistics of Degenerate Bose Systems

FRANZ MOHLING

UNIVERSITY OF COLORADO
Boulder, Colorado

1. FREE BOSE GAS

Historically, the first-known example of a Bose system was the photon gas. In 1924, S.N. Bose[1] first showed that one could understand Planck's equation for the energy-density of black-body radiation from a "quantum-statistical" point of view. His basic assumptions were that many light quanta could occupy the same quantum-mechanical element h^3 of phase space and that the light quanta are indistinguishable. This led him, with the aid of a simple statistical mechanical argument, to the important result for the average number of quanta in a single state:

$$\langle N_i \rangle = \frac{\exp(-\beta\omega_i)}{1 - \exp(-\beta\omega_i)} \qquad (1)$$

where $\omega_i = h\nu_i$ is the energy corresponding to the i-th momentum state and $\beta = (kT)^{-1}$. By a "simple state" was meant an element of phase space $\Omega\delta^3 P_i$ which equals h^3 ($\Omega =$ volume of system). This simple formula could then be used to derive the thermodynamic properties of blackbody radiation by elementary means.

The two important assumptions which have entered into the Bose derivation of equation (1) are that more than one light quantum can occupy a single quantum state and that the light quanta must be considered indistinguishable. The applicability of this assumption can be generalized to include the quantum-statistical analysis of any system of Bose quanta, or particles. Thus, a Bose particle is defined to be any particle which can occupy a quantum state independently of

149

how many other (indistinguishable) Bose particles are already in that state. A Fermi particle is one which cannot occupy a quantum state if another (indistinguishable) Fermi particle is already in that state. One refers to this property of elementary particles, atoms, and molecules by saying that they obey either Bose–Einstein statistics or Fermi–Dirac statistics. It can easily be shown that the generalization of equation (1) to the case of free Bose particles is

$$\langle N_i \rangle = \frac{\exp\left[\beta(g - \omega_i)\right]}{1 - \exp\left[\beta(g - \omega_i)\right]} \tag{2}$$

where $\omega_i = (\hbar^2 k_i^2 / 2M)$ and g is the chemical potential, or thermodynamic potential per particle, of the gas. The significant difference between equations (1) and (2) is the presence in (2) of the quantity g, which is zero for light quanta. This quantity enters into the derivation of equation (2) because of the extra constraint that the number of Bose particles (bosons) must be conserved in the system. Of course, this constraint is not present for black-body radiation.

2. BOSE–EINSTEIN CONDENSATION

The ideal Bose gas exhibits an unusual (theoretical) phenomenon, called Bose–Einstein condensation,[2] which can be understood by computing the density n of the gas using equation (2). Now, at very high temperatures, the denominator of the expression for $\langle N_i \rangle$ can be set equal to unity, and one finds the result

$$n = \Omega^{-1} \sum_i \langle N_i \rangle \xrightarrow[\Omega,\, T\to\infty]{} \lambda_T^{-3} e^{\beta g} \tag{3}$$

where the thermal wavelength λ_T is given by

$$\lambda_T = \left(\frac{2\pi\hbar^2}{MkT}\right)^{1/2} \tag{4}$$

Since $n\lambda_T^3 \ll 1$ at high temperatures, one sees from equation (3) that g is quite negative in that region. Moreover, it increases toward zero as the temperature is decreased.

An important temperature region for examination occurs when the chemical potential g approaches zero, i.e., when $n\lambda_T^3 \lesssim 1$. In this case, one must use the full expression (2) to compute the density. One finds[3] that

$$n\lambda_T^3 = 2.612 - 2\sqrt{\pi}\,(-\beta g)^{1/2} + O(\beta g) \tag{5}$$

Clearly, expression (5) becomes meaningless when $n > (2.612)\lambda_T^{-3}$.

In fact, the number 2.612 is the upper limit of the series (5), because $\beta g \leqslant 0$ would lead to negative average values for the lowest momentum states, according to equation (2). It is obvious that higher densities (or lower temperatures) than those allowed by the limit $n > (2.612)$ $\times \lambda_T^{-3}$ should be possible for the ideal Bose gas. Therefore, equation (5) presents an inconsistency. (This inconsistency does not occur for the photon gas, because the photon density is not an independent variable.)

The resolution to the above inconsistency is achieved by observing that one may isolate a single term in the sum (3), before going to the limit $\Omega \to \infty$ and converting the sum to an integral. The single term is, of course, the number of bosons in the lowest quantum state. One then obtains instead of equation (5)

$$n = n_0 + (2.612)\,\lambda_T^{-3} \qquad \text{for } T < T_c \tag{6}$$

where the critical temperature T_c, for a given density of the ideal Bose gas, is defined by

$$n\left(\frac{2\pi\hbar}{MkT_c}\right)^{3/2} = 2.612 \tag{7}$$

The density n_0 of zero-momentum particles below the critical temperature is found, by solving equations (6) and (7) together, to be

$$n_0 = n\left[1 - \left(\frac{T}{T_c}\right)^{3/2}\right] \tag{8}$$

From the above analysis, one can see that for the ideal Bose gas one is forced to consider the "condensation" of bosons into the zero-momentum state below a critical temperature determined by equation (7). There is, therefore, a macroscopic occupation of the zero-momentum state, so that one can speak of the separation of two phases in equilibrium with each other, not in position space but rather in momentum space. This phenomenon is called Bose–Einstein condensation.

It is not important for the present exposition to go into the full thermodynamic discussion of the ideal Bose gas in the region of the critical temperature.[4] However, it is important to observe that the *momentum space ordering* which occurs in a Bose gas below the critical temperature is a completely quantum-mechanical phenomenon. It occurs in a temperature-density region for which the thermal wavelength λ_T of the bosons is greater than their interparticle spacing $l = n^{-1/3}$. Thus, when $\lambda_T \gtrsim l$, the wave functions of the particles overlap in position space and their Bose statistics cause exchange effects to become very important. The contribution of exchange effects is a major difference between a classical and a quantum mechanical many-body system.

For a real system, one says when $\lambda_T \gtrsim l$, *and* when the kinetic energy of the system is the same order of magnitude (or greater than) the potential energy, that the system is degenerate or strongly degenerate. Degenerate systems can only be understood on a microscopic basis by using the laws of quantum mechanics. They are quantum mechanical! (Note that the term "degenerate," as used here, has little to do with energy-level degeneracy—it is phase-space degeneracy.)

It is believed that there is only one strongly degenerate Bose system in nature. This is liquid helium. If one computes the critical temperature with equation (7), using the known density of helium ($n = 0.146$ g/cm³), then one finds a critical temperature $T_c = 3.13°K$. This is reasonably close to the observed value of the λ-point in liquid helium, $T_\lambda = 2.18°K$, below which the liquid exhibits its superfluid properties. The correlation is probably fortuitous. Nevertheless, it is now generally believed that the theoretical phenomenon of Bose–Einstein condensation has its manifestation in nature in the formation of liquid helium II.

3. DEGENERATE BOSE SYSTEM WITH INTERACTIONS

There still does not exist a microscopic theory of helium II which explains its basic properties quantitatively, including the transition temperature T_λ. The theory of Feynman[5] has been partially successful in that it qualitatively yields the observed excitation curve of helium II, which was first postulated by Landau.[6] Beyond this, there are model calculations and various hydrodynamic calculations which have only shed limited light on the microscopic behavior of helium II. Of course, there is also the extensive work of London, Landau, and others toward the development of a macroscopic, or thermodynamic, understanding of liquid helium. To a large extent, the role of a microscopic theory is merely to provide a rigorous basis for the models of the macroscopic theories. Thus, one of the most important features of the macroscopic theories of helium II is the two-fluid model, first introduced by Tisza.[7] The microscopic interpretation of this model, for a system at rest, is that the "superfluid" is composed of those bosons which occupy the zero-momentum state (macroscopically), whereas the "normal fluid" is composed of all the remaining bosons in nonzero-momentum states.

This interpretation is based on the ideal Bose gas phenomenon of Bose–Einstein condensation discussed above.

In order to study the detailed aspects of a microscopic theory of liquid helium II, it is useful to consider simpler model systems of interacting bosons. A particularly popular model system in the contemporary literature is the dilute Bose gas of hard spheres, because it is believed that the hard cores of the helium atoms must play an essential role in any correct microscopic theory of helium II.[8] Recently, Lee and Yang[9] have explicitly demonstrated that Bose–Einstein condensation occurs in the dilute gas of hard-sphere bosons. Their calculation has therefore placed the two-fluid model of Tisza on a firmer theoretical basis. Moreover, their thermodynamic results for a model Bose system now provide one important check on the validity of any new theory.

From the preceding paragraph, one can see that there is considerable interest in the theory of a general degenerate Bose fluid and not only in the theory of liquid helium II. Consider now, how the development of a microscopic theory of a general degenerate Bose system can proceed. One first writes down the many-body Hamiltonian, which in the second quantized form is

$$H = \sum_{k} a_k^+ a_k \omega_k + \tfrac{1}{2} \sum_{\substack{k_1 k_2 \\ k_3 k_4}} a_{k_1}^+ a_{k_2}^+ \langle k_1 k_2 | V | k_4 k_3 \rangle a_{k_3} a_{k_4} \qquad (9)$$

where $\langle k_1 k_2 | V | k_4 k_3 \rangle$ is a matrix element in momentum space of the elementary two-particle interaction, and where $\omega_k = \hbar^2 k^2 / 2M$. The quantities a_k and a_k^+ are the annihilation and creation operators, respectively, of the free bosons, and these satisfy the usual commutation relations. One next observes that each of the momentum-state sums includes a contribution from the zero-momentum state, which when considered from the point of view of deviations from the free-particle condition may be macroscopically occupied. In order to investigate the ground state ($T = 0$) of the Bose system, it is therefore necessary to treat the zero-momentum state specially.

The difficulty presented by the zero-momentum state can be most easily seen when attempting to calculate the deviations from the free Bose gas condition by applying many-body perturbation theory in a straightforward manner. One assumes in the first approximation that the operators a_0 and a_0^+ may each be replaced by the number $\langle N_0 \rangle^{1/2}$, and one then separates from the interaction part of the Hamiltonian (9) the resulting diagonal terms. Finally, a detailed study[10] of the perturbation treatment of the off-diagonal interaction terms shows that the perturbation theory for the ground-state energy diverges in the limit $\langle N \rangle, \Omega \to \infty$ (keeping $n = \langle N \rangle / \Omega$ finite). It is the macroscopic occupation of the zero-momentum state, i.e., the fact that $\langle N_0 \rangle \sim \langle N \rangle$, which causes this divergence.

The off-diagonal interaction terms in equation (9) which give the dominant contribution to the ground-state energy are those for which both of the annihilation (or creation) operators are $\mathbf{a}_0$ (or $\mathbf{a}_0^+$). These "leading" off-diagonal terms, which involve the momentum-state pairs $(\mathbf{k}, -\mathbf{k})$, can be incorporated into the diagonal part of H by a transformation first introduced by Bogoliubov.[11] This transformation consists of introducing two new annihilation and creation operators $\boldsymbol{\xi}_\mathbf{k}$ and $\boldsymbol{\xi}_\mathbf{k}^+$ by the equations

$$\boldsymbol{\xi}_\mathbf{k} = (1 - \alpha_\mathbf{k}^2)^{-1/2} (\mathbf{a}_\mathbf{k} + \alpha_\mathbf{k} \mathbf{a}_{-\mathbf{k}})$$
$$\boldsymbol{\xi}_\mathbf{k}^+ = (1 - \alpha_\mathbf{k}^2)^{-1/2} (\mathbf{a}_\mathbf{k}^+ + \alpha_\mathbf{k} \mathbf{a}_{-\mathbf{k}}) \tag{10}$$

where $\alpha_\mathbf{k}$ is a complex number which depends only on the magnitude of $\mathbf{k}$ and is determined by the condition of diagonalization. By using this procedure, one finds that the dominant off-diagonal terms of the interaction Hamiltonian result (after diagonalization) in the expression for the ground-state energy per particle

$$\frac{\langle E_0 \rangle}{\langle N \rangle} = 4\pi n a_s' \left(\frac{\hbar^2}{2M}\right) \left\{ 1 - \left(\frac{128}{15}\right) \left[\frac{n(a_s')^3}{\pi}\right]^{1/2} + O[n(a_s')^3] \right\} \tag{11}$$

where

$$a_s' = \left(\frac{4\pi \hbar^2}{M}\right)^{-1} \int d^3r \, V(\mathbf{r}) \tag{12}$$

for a local potential such that $a_s' > 0$.

The result of equations (11) and (12) becomes meaningless when one considers its application to a low-density gas of real Bose particles. The reason is that real particles always have a repulsive core, and a core interaction produces an infinite number for a_s'. The obvious solution to this difficulty is to replace a_s' in equation (11) by the two-particle scattering length a_s, a suggestion due to Landau.[11] Actually, the first explicit calculation of the result (11) was performed by the Lee and Yang[12] for the dilute hard-sphere Bose gas (characterized by $na^3 \ll 1$), in which case the scattering length a_s is simply the diameter a of the hard spheres. Lee, Huang, and Yang[10, 13] have shown how to obtain the result (11) and (12) with the aid of the pseudopotential method, and this method (which cannot be directly applied to a bound system) yields equation (11) with a_s' replaced by a_s. Further discussion of these various methods has been given by Lieb.[14] It may be fairly said that none of the existing methods for deriving the result (11) has a straightforward generalization to the real and important degenerate Bose fluid, helium II.

4. QUANTUM-STATISTICAL THEORY OF THE DEGENERATE BOSE SYSTEM

Perhaps the most promising starting point for the development of a microscopic theory of helium II is the use of the grand partition function Z_G of quantum-statistical mechanics. This quantity can be written as

$$Z_G = e^{\Omega f} = \sum_{N=0}^{\infty} (e^{\beta g})^N \, \mathrm{Tr}_N (e^{-\beta H}) \tag{13}$$

where the symbol Tr_N indicates that the trace of $\exp(-\beta H)$ is to be taken over a complete set of N-particle state vectors. The many-body Hamiltonian H is given in the Fock, or number, representation by equation (9), in which V is the helium two-particle interatomic potential. The quantity f is called the grand potential, and it is an intensive quantity, i.e., the limit of f as $\Omega \to \infty$ exists.

In the application of equation (13) to a general degenerate Bose system, one finds that the ordinary Ursell method of analysis completely fails to yield a well-defined expression for the grand potential. That is, the grand potential, which equals the pressure divided by (kT), seems not to be an intensive quantity, as it should be. Lee and Yang[15] have shown that the reason for this failure is the macroscopic occupation of the zero-momentum state (in a system at rest). They observe that if L is the number of bosons in the zero-momentum state in any given term of the trace in equation (13), then each of the $L!$ (quantum-mechanical) exchange terms corresponding to this given term gives an identical contribution. It is the proliferation of identical exchange terms when $\langle L \rangle \sim \langle N \rangle \gg 1$ that causes the breakdown of the usual Ursell method for determining the grand potential.

To be sure, it is this same proliferation of exchange terms which causes Bose-Einstein condensation, as indicated below equation (8). There is, therefore, not necessarily anything wrong with equation (13) for the grand partition function. Rather, attention is focused on one's inability to apply equation (13) to the microscopic understanding of a degenerate Bose system.

Lee and Yang[15] have also shown how to overcome the difficulty presented by equation (13) with their x-ensemble formulation of the grand partition function. In this formulation, the use of the grand canonical ensemble includes the introduction of a parameter x which is the density of zero-momentum particles in the degenerate Bose system [$\langle x \rangle = n_0$ of equation (8)]. Thus, in the "x-ensemble" formulation, the grand partition function is given by

$$e^{\Omega f_x} = e^{-\Omega x} \sum_{N=0}^{\infty} (e^{\beta g})^N \sum_{L=0}^{N} \frac{(x\Omega)^L}{L!} \operatorname{Tr}_{L,N}(e^{-\beta H}) \qquad (14)$$

instead of by equation (13). In this new expression $\operatorname{Tr}_{N,L}$ means that the trace is to be taken over those N-particle state vectors in which L (and only L) particles have zero momentum. In their proof of equation (14), Lee and Yang have shown that if

$$\frac{\partial f_x}{\partial x} = 0 \quad \text{at} \quad x = \langle x \rangle = \frac{\langle L \rangle}{\Omega} > 0 \qquad (15)$$

then $f_x = f$ in the limit $\Omega \to \infty$. If equation (15) is not satisfied, then $f_x = f$ at $\langle x \rangle = 0$, and equation (14) reduces to equation (13), for all practical purposes. It can be seen that in the x-ensemble formulation the macroscopic occupation of a single quantum state ($\mathbf{k} = 0$) is allowed from the very beginning, although it is not required that $\langle x \rangle > 0$. The use of equation (14) for the grand partition function is somewhat analogous to the replacement of the operators a_0 and a_0^+ by $\langle N_0 \rangle^{1/2}$ in the many-body perturbation theory discussed below equation (9).

It is easy to show how the x-ensemble expression (14) is obtained from equation (13), although the following discussion does not constitute a formal proof of equations (14) and (15). The motivating step is to eliminate the troublesome factor $(L!)$ from the $\operatorname{Tr}_{N,L}$ in equation (13). One next observes that when the average number $\langle L \rangle$ of zero-momentum bosons is $\sim \langle N \rangle$, then the dominant terms in the grand partition function must be those with $L \sim \langle L \rangle$. One can see this by making an *a posteriori* argument to show that the fluctuation $\langle (\Delta L)^2 \rangle$ in the number of zero-momentum particles, calculated by using equation (14), is $\langle L \rangle$. Then $\langle L \rangle^{-1} \langle (\Delta L)^2 \rangle^{1/2} = \langle L \rangle^{-1/2} \lll 1$ for $\langle L \rangle \ggg 1$.

One next multiplies the $\operatorname{Tr}_{N,L}$ in the grand partition function (13) by the ratio $(\langle L \rangle !)(L!)^{-1}$, which is essentially unity for important L values, and this eliminates the troublesome $(L!)$ factor. To arrive at equation (14), one has finally to use Stirling's approximation for $\langle L \rangle !$, a step which is valid because $\langle L \rangle \ggg 1$, and then to replace $\langle L \rangle = \langle x \rangle \Omega$

by $x\Omega = L$. In this final step, the density x is treated as a variable whose average value can be obtained by the maximum condition (15). This step is necessary in order that one can have a method for calculating $\langle x \rangle$.

With the grand partition function (14), one can apply the Ursell method to arrive at an explicit expression for the grand potential f_x. It is not the purpose of the present discussion to go through the details of this subsequent analysis; it has been carried through in a straight-forward manner by Lee and Yang[15] and, more recently, by Mohling.[16] For the free Bose gas, it is fairly easy to derive from equation (14) the expression

$$f_x = -\Omega^{-1} \sum_{\rho \neq 0} \ln \left(1 - e^{\beta(g - \omega_p)}\right) - x + x e^{\beta g} \tag{16}$$

This agrees with the expression which one would derive from equation (13), because when $\langle x \rangle > 0$, i.e., when $T < T_c$, then $g = 0$ [see equation (5) and subsequent discussion].

In order to how clarify equation (14) is to be used in an actual cal-culation, it is useful to exhibit the relation between the grand potent-ial and various thermodynamic quantities. Treating β, g, and x as independent variables, one can derive from equation (13) the fol-lowing expressions:

Pressure

$$P = \beta^{-1} \frac{\partial}{\partial \Omega} (\Omega f) \tag{17}$$

Particle density

$$n = \Omega^{-1} \langle N \rangle = \beta^{-1} \frac{\partial f}{\partial g} \tag{18}$$

Energy per particle

$$\frac{\langle E \rangle}{\langle N \rangle} = g - n^{-1} \frac{\partial f}{\partial \beta} \tag{19}$$

Entropy

$$S = \frac{\partial}{\partial T} (\beta^{-1} \Omega f) \tag{20}$$

Upon subsituting equation (16) into equation (18), one finds for the density of the free Bose gas

$$n = \Omega^{-1} \sum_{\rho \neq 0} \left[\frac{e^{\beta(g - \omega_p)}}{1 - e^{\beta(g - \omega_p)}} \right] + x e^{\beta g} \tag{21}$$

This agrees with equations (6) and (7), because when $\langle x \rangle > 0$, then equation (15) yields for the free Bose gas

$$\frac{\partial f_x}{\partial x} = 0 = -1 + e^{\beta g} \tag{22}$$

which implies that $g = 0$, as pointed out below equation (16). It is important to observe that the determination of $\langle x \rangle$ when $\langle x \rangle > 0$ has required in this case the use of both equations (15) and (18), and not merely equation (15) alone; this is quite generally true. Other thermodynamic expressions for the free Bose gas can readily be derived from equations (16), (17), (19), and (20).

As a concluding remark, it is of interest to observe that equation (14) can also be obtained from equation (13) by arguing that the factor $e^{-\Omega x}(x\Omega)^L (L!)^{-1}$ is just the Poisson distribution of L about the average value $(x\Omega)$. Since $\langle x \rangle \Omega \gg 1$, and the fluctuations in L are small, equation (13) will be essentially unchanged if each term $\mathrm{Tr}_{N,L}$ ( ) in the sum over all N and L is multiplied by this Poisson distribution function. This observation was made by Prof. L. Rosenfeld, Chairman of the Symposium.

REFERENCES

1. S.N. Bose, *Ztschr. Phys.* **26**: 178 (1924); **27**: 384 (1924).
2. A. Einstein, *Preuss. Akad. Wiss. (Berlin)* **22**: 261 (1925); **1**: 3 (1925).
3. F. London, *Superfluids, Vol. II*, John Wiley and Sons, New York, 1954. (Appendix.)
4. F. London, *Superfluids, Vol. II*, John Wiley and Sons, New York, 1954, pp. 40–53.
5. R.P. Feynman, *Phys. Rev.* **94**: 262 (1954).
6. L. Landau, *J. Phys. (USSR)* **11**: 91 (1947); **5**: 71 (1941); **8**: 1 (1944).
7. L. Tisza, *Nature* **141**: 913 (1938); *Phys. Rev.* **72**: 838 (1947).
8. F. London, *Superfluids, Vol. II*, John Wiley and Sons, New York, 1954, pp. 30–35.
9. T.D. Lee and C.N. Yang, *Phys. Rev.* **112**: 1419 (1958)..
10. T.D. Lee, K. Huang, and C.N. Yang, *Phys. Rev.* **106**: 1135 (1957).
11. N.N. Bogoliubov, *J. Phys. (USSR)* **11**: 23 (1947).
12. T.D. Lee and C.N. Yang, *Phys. Rev.* **105**: 1119 (1957).
13. T.T. Wu, *Phys. Rev.* **115**: 1390 (1959).
14. E.H. Lieb, *Phys. Rev.* **130**: 2518 (1963).
15. T.D. Lee and C.N. Yang, *Phys. Rev.* **117**: 897 (1960).
16. F. Mohling, "Degenerate Bose System. I. Quantum Statistical Theory," *Phys. Rev.* **135**: A 831 (1964).

Some Current Trends in Mathematical Research

M. H. STONE*

UNIVERSITY OF CHICAGO
Chicago, Illinois

One of the most striking features of modern mathematics is the high degree of specialization of the individual researcher. The result of this is that workers in one area are apt to be totally ignorant of many other areas (and sometimes even proud of this fact). The mathematician's attitude to other disciplines such as physics and economics—in which the workers are interested not in mathematics *per se* but only in its applications to their problems—is something like: "I try to help everyone, but I have also my own concerns."

When compared to other sciences, mathematics appears to progress in a less erratic fashion, on more fronts at once, and in a fairly steady way. Almost any branch of mathematics that has ever been worked on is still being worked on now, and none tends to predominate over another. On the other hand, in physics, for example, this may not be so: We find that in successive periods of time, relativity theory, quantum theory, quantum field theory, or the theory of elementary particles was the predominant field of interest. One reason for this would appear to be the fact that experimental necessities, involving the investment of huge sums of money, forced concentration on one particular area of the subject at a time. Further, the study of physics has a definite objective, namely, the understanding of nature, which justifies such a procedure, whereas mathematics is perhaps not so single-minded a discipline.

We now single out a few major areas of mathematics and survey the research trends therein.

---

* Distinguished Service Professor of Mathematics.

1. NUMBER THEORY

This is one of the oldest branches of mathematics, starting with simple results on divisibility and prime numbers, proved by simple methods. Later came the prime number theorem (How many primes are there between 1 and n, where n is an arbitrary natural number?), the statement of which was arrived at from empirical considerations and then proved rigorously. The German mathematician Riemann introduced an analytical approach to number-theory problems through his zeta-function; this approach had many side-effects in analysis, but it fell short of its goal with respect to number theory. The current trend is to use other methods involving more elementary functions, especially the exponential and the logarithmic functions, and relying on simpler arguments. It may be said that this area currently engages comparatively few prominent mathematicians, in marked contrast to what was true earlier in the century.

2. CONFORMAL MAPPING THEORY

Since its inception (by Riemann), this theory has developed in technical ways; one study has taken the form of the preparation of an atlas of conformal maps, and another concerns itself with conformal maps between Riemann surfaces and the classification of these surfaces. The topological characterization of conformal maps gives rise to analytical topology.

In extending the one-variable theory to the theory of functions of several complex variables, serious difficulties were encountered. Only from the 1930's onward has substantial work been done. The most fruitful work now involves the topological and differential-geometric properties of complex manifolds (surfaces which behave locally like the product of complex planes and have the necessary differential characteristics). The one-variable theory, as is well known, gives rise to certain partial differential equations, the Cauchy–Riemann and the Laplace equations. One pretty theory (due to Bers and Ahlfors) involves the generalization of the Cauchy–Riemann equations to the case of nonconstant coefficients.

3. DIFFERENTIAL EQUATIONS

This branch is as old as mechanics. There is a classical partitioning of the subject into two topics—ordinary differential equations and partial differential equations. Some of the interest in this subject derives from practical applications, especially in the case of nonlinear ordinary differential equations. In partial differential equation theory, there appears to be a renewed interest, thanks to the problems posed by studies in plasticity, nonlinear hydrodynamics, and turbulence theory. In the case of linear differential equations, recent work on functional analysis gives us powerful tools to tackle existence theorems and the proper setting of boundary questions; some progress has been made even in the nonlinear case.

But, on the whole, one is left with the feeling that our knowledge of differential equations is highly inadequate, and even the most elementary types of partial differential equations baffle analysis. Partial differential equations outside the three classic types (the elliptic, parabolic, and hyperbolic) do not have physical significance. The problem becomes that of asking the right questions about differential equations, and perhaps geometry will be the best guide here.

4. GROUP THEORY

This branch of mathematics was in existence even before the technical term "group" was introduced. Two major areas of the subject are:

1. *Finite Groups.* These provided the source of group theory as a theory; in the last few years, some long-standing problems in this area have been solved, for example, the conjecture that all odd-order groups are solvable.

2. *Continuous Groups.* This includes the study of Lie groups and Lie algebras; we still do not have adequate knowledge as to how to study the group representations, even of Lie groups; the general solution of the problem of representation of groups in linear spaces is still

essentially incomplete. However, important special cases have been thoroughly treated—for example, the unitary representations in Hilbert space.

5. DIFFERENTIAL GEOMETRY

This had its origin with Gauss and Riemann, though to some extent they were anticipated by Euler and others in this work. It has had a steady development throughout the last century and the beginning of the present; in particular, we may mention the development of the tensor calculus. But, until about 1950, the subject had a sort of fixed form and methodology. Then a new point of view was introduced, establishing connections between general topological spaces and differentiable manifolds (hypersurfaces). A good reason for such a development was the readily available knowledge of the topology of hypersurfaces. This led to a re-examination of the foundations of differential geometry and of the questions to be asked in the subject. In the last twelve years or so, extraordinary progress has been made. We may cite one of the most remarkable results in this area: On certain hyperspheres, it is possible to introduce two nonequivalent differentiable structures. The whole story is still far from being known. It appears difficult to guess at the practical implications of this development, although one may have a hunch that analytical applications may prove to be more important than topological ones.

NEW AREAS

Apart from the above areas having a long history, the present century has seen the birth of quite a few new branches of mathematics. We shall discuss two of them below.

Mathematical Logic

It was the dream of Leibnitz that mathematics could be used in formal logic and that a calculus could be set up for such use. We may trace the origin of the subject itself to Boole's "Laws of Thought," published around 1850. Important contributions were made by Russell and Whitehead, who set up a symbolic system expressing the principles of logic and including set theory and the function concept.

Let us give a brief description of the method of symbolic logic. Just as in the formation of words and sentences in ordinary speech and writing, we consider finite sequences of certain objects and prescribe a rule for combining such sequences ("concatenation"); thus, we obtain a "monad" with the above sequences as elements. By manipulations of this monad, every type of algebra can be constructed, and logic itself can be reduced to the algebra of a special kind. This way of looking at logic is intimately connected with modern ideas of computation and the associated concepts of the speed, capacity, and memory of a machine.

Studies in this area led to certain disturbing features of the logic used in mathematical proofs; one of the most celebrated is the Russell paradox, whose immediate moral was that making "self-references" should be avoided. This obviously not being rigorously possible, the next problem was one of choosing between what self-references to admit and what to avoid: the difficulty now is that there is no *objective* rule for deciding this question.

Another troublesome question was that of the "consistency" of mathematics. One of Hilbert's famous problems, posed in 1900, concerned itself with this question; for twenty years after this, the problem lay untouched, and then Hilbert himself took it up. By 1931, Göden's incompleteness theorem (accompanied by some completeness theorems also, by way of softening the blow) revealed the futility of such efforts.

Mathematical logic is currently a very active area of interest. A recent result of Cohen has evoked much attention; it establishes the independence of the continuum hypothesis.

Topology and Algebra

The study of topology had its beginnings in problems such as the structure of the Möbius strip and the Königsberg seven-bridge problem. It did not have an existence of its own until Poincaré made it into a separate mathematical discipline, in the latter part of the nineteenth century; he was concerned mostly with a matrix formulation. By about 1920, American and Russian mathematicians began working in the direction pointed out by Poincaré, and made topology into a tremendously active field. It has contributed to the study of differential equations, but is not very intimately connected to other branches of mathematics used in the applications.

Since linear algebras are linked to topology, topology has come to have a strong algebraic aspect; at the same time, continuity considerations have imparted to it an analytical aspect as well. A related development was that of the subject of homological algebra.

In the same period as that of the development of topology, algebra itself made tremendous strides. About 1920, Dickson and others introduced essentially new approaches. In Europe, Emmy Noether and her disciples, especially Van der Waerden and Artin, made important contributions. These developments had their impact on algebraic geometry (the study of systems of polynomials), which was transmuted into a purely algebraic discussion, a development which was unsuccessfully resisted by the Italian school of geometers.

We now turn to a summing up of the broad aspects of these developments. As already stated, after 1920, algebra became highly influential in algebraic geometry and in topology. Functional analysis provides another example of a subject which is highly algebraic in character, and is a natural field of study for the physicist, especially one interested in quantum theory. The methods involved are a combination of algebraic, geometric, and topological ones. One is inevitably struck by the unifying algebraic and/or topological features of many branches of modern mathematics. A very curious phenomenon, probably temporary, is that all the time we deal with only a small number of basically different mathematical systems, the rest being merely combinations thereof. On the one hand, we have the algebraic systems (monads, dyads, etc.), and on the other the ordered systems (both simply ordered and partially ordered), which are essentially the topological systems. (Incidentally, we note that almost all topological systems are based on the real number system; exceptions, however, do exist.)

At the present time, the tendency is to go from one level of abstraction to a higher one, and away from specific problems to problems on general systems, i.e., to comparative studies of structure. For example, in 1938–1939, the concept of category was introduced by Eilenberg and Maclane. It permits the development of an abstract theory of comparison. The result is that the difference between abstract and concrete tends to disappear, the "abstract" of today becoming the "concrete" of tomorrow.

Semigroup Methods in Mathematical Physics

A. T. Bharucha-Reid*

WAYNE STATE UNIVERSITY
Detroit, Michigan

---

1. INTRODUCTION

The theory of one-parameter semigroups of operators in Banach spaces is concerned with the study of the operator-valued exponential function, its representation and properties, in infinite-dimensional function spaces. The theory can also be considered as a generalization of Stone's theorem[33] on the representation of one-parameter groups of operators in Hilbert space. The theory of semigroups of operators had its origin in 1936, when Hille[10] investigated certain concrete semigroups of operators. In 1948, Hille[11] published his treatise on semigroups of operators; and in the 1950's this treatise stimulated a considerable amount of research on semigroups and their applications. In 1957, Hille and Phillips[12] published an extensive revision of Hille's treatise, and at the present time their treatise is the bible of workers in semigroup theory. In addition to the Hille–Phillips treatise, semigroups of operators are the subject of lecture notes by the author[2] and Yosida[40]; chapters on semigroups of operators can be found in the books of Dunford and Schwartz,[5] Maurin,[20] and Riesz and Sz.-Nagy.[30] We also refer to the papers of Phillips,[27, 29] the first of which presents a survey of semigroup theory, while the second gives a very readable introduction to semigroup theory and its applications in the theory of partial differential equations.

To introduce the semigroup relation, we proceed as follows: Let $f(t)$ be a real function of t, $t \geqslant 0$. A classical problem is to find the most general continuous function $f(t)$ which satisfies the functional equation

---

* Visiting Professor, Faculty of Applied Mathematics, Institute of Mathematical Sciences, 1963–1964.

$$f(s + t) = f(s)f(t) \qquad s, t > 0 \tag{1.1}$$

with $f(0) = 1$. It is well known that the exponential function $f(t) = e^{\lambda t}$, λ a constant, is the most general continuous real function which satisfies equation (1.1).

Let us now consider the operator analog of equation (1.1). Let $\mathscr{X}$ be a Banach space, $T(t)$ a linear operator on $\mathscr{X}$, and let $\mathscr{E}(\mathscr{X})$ denote the Banach algebra of endomorphisms (bounded linear operators) of $\mathscr{X}$. In this case, we wish to determine the most general continuous operator-valued function $T(t)$: $(0, \infty) \to \mathscr{E}(\mathscr{X})$ which satisfies the functional equation

$$T(s + t) = T(s)T(t) \qquad s, t > 0 \tag{1.2}$$

with $T(0) = I$ (the identity operator). A family of operators $\{T(t), t \geq 0\}$ satisfying equation (1.2) is called a *one-parameter semigroup of operators* in the Banach space $\mathscr{X}$. If $T(t)$ admits an inverse $T^{-1}(t) = T(-t)$, so that $T(t) T(-t) = I$, we say that the family of operators $\{T(t), -\infty < t < \infty\}$ forms a *one-parameter group of operators* in the Banach space $\mathscr{X}$.

From the classical case considered above, one might expect that the operator-valued function $T(t)$ satisfying equation (1.2) admits, in some sense, a representation as an exponential function. This is indeed the case; and as it turns out the following two situations arise: (1) if $T(t)$ is continuous in the uniform operator topology, then there exists a bounded operator A on $\mathscr{X}$ such that $T(t) = e^{\cdot tt}$; and (2) if $T(t)$ is continuous in the strong operator topology, then $T(t)$ can be represented, in an appropriate sense, as $e^{\cdot tt}$, but in this case A is an unbounded operator. The operator A is called the infinitesimal generator of the semigroup. (For a detailed discussion of the representation of semigroups, see Bharucha-Reid, chapter 2,[2] and Hille and Phillips, chapter 11.[12])

The purpose of this paper is to present an introductory survey of semigroup methods in mathematical physics. Semigroup methods are, in the main, introduced in mathematical physics by the many initial-value problems that arise in the representation of the evolution of physical systems as partial differential equations. In this connection it is of interest to quote from the recent book by Mackey[19]: "Our fundamental viewpoint is that the change in time of a physical system may be described by a one-parameter semigroup U acting on a set S and that the laws of physics make assertions about the structure of S and the 'infinitesimal generator' of U." The application of semigroup theory to the evolution of physical systems is based on the fact that the solution of the Cauchy problem for a linear partial differential equation, with prescribed initial values at $t = 0$ belonging to a concrete Banach space, is obtained by applying an operator $T(t)$ to the initial values. The family of operators $\{T(t),\ t \geqq 0\}$ forms a semigroup, the associated infinitesimal generator being a differential operator.

As an example, we shall briefly consider the well-known diffusion, or heat, equation

$$\frac{\partial u}{\partial t} = \frac{\partial^2 u}{\partial x^2} \qquad u = u(t, x) \tag{1.3}$$

for $t > 0$, $x \in (-\infty, \infty)$, and satisfying the initial condition

$$\lim_{t \to 0} u(t, x) = f(x) \tag{1.4}$$

We take the Banach space $\mathscr{X}$ in this case to be $C(-\infty, \infty)$—the space of continuous functions on $(-\infty, \infty)$; hence $f(x) \in C(-\infty, \infty)$. To stress the dependence of the solution $u(t, x)$ on the particular initial value $f(x)$, we will denote the solution of equation (1.3) associated with the initial condition (1.4) by $u(t, x; f)$. We now make the following assumption of uniqueness: namely, there exists at most one solution of equation (1.3) for a given x. We also assume that $u(t, x; f) \in C(-\infty, \infty)$ for each $t > 0$. It now follows from the linearity of equations (1.3) and (1.4), and the uniqueness assumptions, that

$$u(t, x; \alpha f + \beta g) = \alpha u(t, x; f) + \beta u(t, x; g) \tag{1.5}$$

Let us now consider how we might obtain a solution $u(s_1 + s_2, x; f)$. Such a solution, giving the state of the system at time $s_1 + s_2$ can be obtained directly from $f(x)$ at time $t = s_1 + s_2$, or it can be obtained from the solution $u(s_2, x; f)$ after time $t = s_1$. Hence, we can write

$$u(s_1 + s_2, x; f) = u[s_1, x; u(s_2, x; f)] \tag{1.6}$$

In view of the above, the solution $u(t, x; f)$ defines a linear operator $T(t)$ on $C(-\infty, \infty)$ to itself as follows:

$$u(t, x; f) = T(t)[f(x)] \tag{1.7}$$

From equations (1.6) and (1.7) we see that

$$T(s_1 + s_2)[f(x)] = T(s_1)T(s_2)[f(x)] \qquad s_1, s_2 > 0 \tag{1.8}$$

The initial condition (1.4) can be expressed in the form

$$\lim_{t \to 0} T(t)[f(x)] = f(x) \qquad f(x) \in C(-\infty, \infty) \tag{1.9}$$

or

$$\lim_{t \to 0} T(t) = I \qquad (1.9')$$

Hence we can conclude that the family of operators $\{T(t), \ t \geqq 0\}$, where $T(t)$ is defined by (1.7), forms a semigroup of operators in $C(-\infty, \infty)$. We will say more about this semigroup in Section 3.

It is of interest at this point to refer to Hadamard's analysis of Huyghen's principle in optics, and the semigroup property implied by equations (1.6) and (1.7). In 1903, Hadamard[9] noted that Cauchy's problem for the wave equation led to certain groups of transformations, and that the group property implied, and was implied by, certain transcendental "addition theorems" satisfied by the elementary solutions used to construct the solution of the wave equation. We remark that the wave equation, which describes reversible phenomena, is of hyperbolic type and leads to a group of operators, while the diffusion equation, describing irreversible phenomena, is of parabolic type and leads to a semigroup of operators.

Hadamard found that the group property of the transformations was a consequence of what he termed the *principle of scientific determinism*. This principle can be stated as follows:

> Given the state of a physical system at time $t = t_0$ $(t_0 \geqq 0)$, one can deduce the state of the system at time $t > t_0$.

An important corollary of this principle was formulated by Hadamard, and is referred to as the *major premise of Huyghen's principle*. This corollary can be stated as follows:

> The state of a physical system at time t $(t > 0)$ can be deduced from its state at an intermediate time t_1 by first computing the state at time t_1 and then from the latter the state at time t, the result being the same as that obtained by direct computation from the original state at time t_0.

In mathematical form, the major premise of Huyghen's principle is expressed by (1.6), (1.7), and (1.8).

The remainder of this paper is divided into four sections. In Section 2 we state some definitions and theorems of semigroup theory. The theorems will be presented without proof, since the main purpose of this section is to give a list of the definitions and results that will be used or referred to in the subsequent sections. Section 3 is devoted to a discussion, from the point of view of semigroup theory, of two well-known differential equations of mathematical physics. In Section 4 we consider the semigroups of operators which arise in quantum mechanics. The last section, Section 5, is concerned with semigroups as solutions of the Boltzmann equation of transport theory.

While this paper was being prepared, three interesting papers appeared which should be studied by any reader seriously interested in semigroup methods in mathematical physics. Two of the papers, by Feldman[7] and Nelson,[24] are concerned with the Schrödinger equation. The third paper, by Segal,[32] considers *nonlinear* semigroups and their applications to certain problems in relativistic physics. We regret that time did not permit us to include a discussion of these studies in this survey.

2. SEMIGROUP THEORY: SOME DEFINITIONS AND THEOREMS

A. Introduction

In this section, we present an outline of the elements of semigroup theory. Although we shall restrict our attention to the definitions and theorems which will be used in the subsequent sections, we hope that this outline will serve as a useful introduction to the theory of semigroups of operators. For proofs of the theorems stated in this section, and for a detailed discussion of semigroup theory, we refer to the works of Bharucha-Reid,[2] Hille and Phillips,[12] Yosida,[40] and other cited references.

B. Fundamental Operators of Semigroup Theory

There are three fundamental operators of semigroup theory, namely, the *semigroup operator*, the *infinitesimal operator* (or *generator*) of the semigroup, and the *resolvent* of the infinitesimal operator.

Definition 1. A family $\{T(t), t \geqq 0\}$ of endomorphisms in a Banach space $\mathscr{X}$ is called a *one-parameter strongly continuous semigroup of operators* if the following conditions are satisfied:

(A) $T(s + t) = T(s) T(t)$, where $s, t \in (0, \infty)$.

(B) $T(0) = I$.

(C) For each $x \in \mathscr{X}$, $T(t)x$ is strongly continuous in t for $t > 0$.

If, in addition, the mapping $T(t): (0, \infty) \to \mathscr{E}(\mathscr{X})$ is continuous in the uniform operator topology, then $\{T(t), t \geqq 0\}$ is called a *uniformly continuous semigroup of operators*. Throughout this paper the term *semigroup operator* will refer to an operator $T(t)$, $t > 0$, which belongs to a family of operators $\{T(t), t \geq 0\}$ with the semigroup property.

We now define two important types of semigroups which are frequently encountered in applications:

Definition 2. A semigroup $\{T(t), t \geq 0\}$ such that $\|T(t)\| \leq 1$ is called a *semigroup of contraction operators* or simply a *contraction semigroup*. A semigroup of (positive) contraction operators with the property that $\|T(t) x\| = \|x\|$ for all $x \in \mathscr{X}^+$ (the positive cone of $\mathscr{X}$) is called a *semigroup of transition operators* or simply a *transition semigroup*.

Semigroups can be classified on the basis of the convergence of the semigroup operator $T(t)$ as $t \to 0$. This classification is based on the sense, or the type, of convergence. There are six basic classes of semigroups; however, we will consider only one of these classes.

Definition 3. A semigroup $\{T(t), t \geq 0\}$ is said to be of *class* (C_0) if $\lim_{t \to 0} T(t) x = x$, the limit being taken in the strong sense.

The next operator to be introduced plays a central role in semigroup theory. The operator we now consider is called the infinitesimal operator, and, as we shall see, this operator reflects the infinitesimal properties of the semigroup operator; and in terms of this operator the semigroup admits various representations.

Definition 4. Let

$$A_\xi = \frac{1}{\xi} [T(t) - I] \qquad \xi > 0$$

The *infinitesimal operator* of the semigroup $\{T(t), t > 0\}$ is the linear operator A_0 defined by

$$A_0 x = \lim_{\xi \to 0} A_\xi x, x \in \mathscr{X}$$

whenever the (strong) limit exists. The least closed extension of A_0, when it exists, is called the infinitesimal generator of the semigroup, and will be denoted by A.

For semigroups of class (C_0), $A_0 = A$. Since we are restricting our attention to semigroups of class (C_0), we will from now on refer to the infinitesimal generator of a semigroup rather than its infinitesimal operator.

Definition 5. The set of elements $x \in \mathscr{X}$ for which $\lim_{\xi \to 0} A_\xi x$ exists is called the *domain* of A, which we denote by $\mathscr{D}(A)$.

It is clear that A is a linear operator, and that $\mathscr{D}(A)$ is a linear subspace of $\mathscr{X}$. We also remark that $\mathscr{D}(A)$ is dense in $\mathscr{X}$.

A basic theorem concerning the differentiability of semigroups is as follows:

Theorem 1:   If the semigroup operator $T(t)$ is strongly continuous for $t > 0$, then for $x \in \mathscr{D}(A)$ the following differential equations are satisfied:

$$\frac{dT(t)x}{dt} = AT(t)x = T(t)Ax \qquad t > 0$$

Finally, we state the following theorem.

Theorem 2:   If $\{T(t), t \geq 0\}$ is a semigroup of class (C_0), then A is a closed linear operator.

The resolvent of an operator L is defined as $R(\lambda; L) = (\lambda - L)^{-1}$, and the values of the (complex) parameter λ for which $R(\lambda; L)$ is defined i.e., the values of λ for which $(\lambda - L)$ has a bounded inverse, form the *resolvent set* $\rho(L)$ of L. The complement of $\rho(L)$ is called the *spectrum* $\sigma(L)$ of L. We now introduce the resolvent operator of semigroup theory, restricting our attention to strongly continuous semigroups of class (C_0).

Definition 6.   The Laplace integral

$$R(\lambda; A)x = \int_0^\infty e^{-\lambda \xi} T(\xi)x \, d\xi$$

which converges for Re $(\lambda) > \omega_0$ (where $\omega_0 = \lim_{t \to \infty} (1/t) \log \|T(t)\|$), regardless of $x \in \mathscr{X}$, is called the *resolvent operator*.

From the above definition, we see that $R(\lambda; A)$ is the abstract Laplace transform of the semigroup operator $T(t)$; hence, it is reasonable to assume that the semigroup operator can be obtained by inversion from the resolvent operator. This is indeed the case; and we now state the following inversion theorem:

Theorem 3:   For every $x \in \mathscr{X}$, $t \geq 0$, and for $\gamma = \max(0, \omega_0)$

$$\int_0^\infty T(\xi)x \, d\xi = \lim_{\omega \to \infty} \frac{1}{2\pi i} \int_{\gamma - i\omega}^{\gamma + i\omega} e^{\lambda t} R(\lambda; A) \, x \, \frac{d\lambda}{\lambda}$$

the limit existing uniformly with respect to t in any finite interval. For every $x \in \mathscr{D}(A)$, the semigroup operator $T(t)$, $t > 0$ admits the representation

$$T(t)x = \lim_{\omega \to \infty} \frac{1}{2\pi i} \int_{\gamma - i\omega}^{\gamma + i\omega} e^{\lambda t} R(\lambda; A)x \, d\lambda$$

with the limit existing uniformly with respect to t in any interval $(\mathscr{E}, 1/\mathscr{E})$, $\mathscr{E} > 0$.

The above theorem, which is important in various applications, enables us to obtain the semigroup operator if the resolvent operator is known.

C. Generation of Semigroups

Because of the central role played by the infinitesimal generator in semigroup theory, it is clear that a very important problem is that of determining when a given closed linear operator is the infinitesimal generator of a semigroup of operators. This problem is of greatest importance in applied problems; for in these cases we often derive an "infinitesimal generator" on the basis of a given physical model. It is therefore of great interest to know what conditions that operator must satisfy in order to be the infinitesimal generator of a semigroup, since it is the semigroup operator which enters as a "solution operator" in many applied problems.

The basic generation theorem for semigroups of class (C_0) which we now state is due to Feller, Miyadera, and Phillips[12]:

Theorem 4: A necessary and sufficient condition that a closed linear operator A on a Banach space $\mathscr{X}$ generate a semigroup $\{T(t), t \geq 0\}$ of class (C_0) such that $\|T(t)\| \leq M$ (M a constant) is that $\mathscr{D}(A)$ be dense in $\mathscr{X}$, and $\|[R(\lambda; A)]^n\| \leq M\lambda^{-n}$ for $\lambda > 0$ and $n = 1, 2, \ldots$.

For applications, the above generation theorem has the disadvantage of requiring a bound on all the positive integral powers of the resolvent operator. A corollary of the Feller–Miyadera–Phillips theorem is the celebrated Hille–Yosida theorem, which for many years was "the" generation theorem. This theorem can be stated as follows:

Theorem 5: If A is a closed linear operator with dense domain, if the resolvent operator $R(\lambda; A)$ exists for $\lambda > 0$, and if $\|R(\lambda; A)\| \leq 1/\lambda, \lambda > 0$, then A is the infinitesimal generator of a semigroup of contraction operators of class (C_0).

Since the Hille–Yosida theorem requires a bound on only the first power of the resolvent operator, it is this theorem which continues to play a central role in the study of concrete semigroups arising in applied problems.

The results stated above hold when the space $\mathscr{X}$ is an arbitrary Banach space, so it might be expected that simpler criteria hold in the case where $\mathscr{X}$ is a Hilbert space. This is indeed so, and we now give two theorems, due to Phillips,[28] on the generation of semigroups in Hilbert space. Because many problems in mathematical physics have their setting in Hilbert space, the theorems we now state are applicable to those problems which utilize semigroup theory.

We first introduce the following definition:

Definition 7. An operator A on a Hilbert space H is said to be *dissipative* if

$$(Ax, x) + (x, Ax) \leqq 0, x \in \mathscr{D}(A)$$

and A is said to be maximally dissipative if it is not the proper restriction of any other dissipative operator.

A necessary and sufficient condition that an operator A on a Hilbert space H is maximally dissipative is given by the following result: Let $\lambda > 0$, and suppose that A is a dissipative operator with dense domain. Then A is maximally dissipative if and only if $\mathscr{R}(\lambda - A) = H$, i.e., the range of $\lambda - A$ is the space H.

We now state the following generation theorem:

Theorem 6: A necessary and sufficient condition for an operator A to generate a strongly continuous contraction semigroup in a Hilbert space is that A be maximally dissipative with dense domain.

A particular subclass of contraction operators of interest in quantum mechanics is the class of isometric operators or isometries. In dealing with isometric operators, we need the notion of a conservative operator defined as follows:

Definition 8. An operator A on a Hilbert space is called conservative if $(Ax, x) + (x, Ax) = 0$, $x \in \mathscr{D}(A)$.

Before stating the generation theorem for semigroups of isometric operators, we give the following results: An operator A is conservative if and only if iA is symmetric. If A is conservative and maximally dissipative with dense domain, then iA is maximally symmetric.

Theorem 7: A necessary and sufficient condition that an operator A generate a semigroup of isometric operators in a Hilbert space is that A be conservative and maximally dissipative with dense domain.

We close this section on the generation problem by remarking that the infinitesimal generator of a semigroup is unique; that is, a closed linear operator can be the infinitesimal generator of at most one semigroup of class C_0. This result follows from the uniqueness theorem for Laplace transforms.

D. Some Additional Topics

In this section we consider some additional topics in semigroup theory which we feel might be of interest to workers in mathematical physics. We first consider the perturbation of semigroups of class C_0. Next, we give a result on the representation of contraction semigroups in Hilbert space. Finally, we introduce the notion of equivalent semi-groups.

Perturbation theory has long been a subject of interest to analysts and applied mathematicians, and in certain branches of mathematical physics, perturbation techniques are employed as a standard tool. The first theorem we state gives a series representation for the per-turbed semigroup:

Theorem 8: If A is the infinitesimal generator of a semigroup $\{T(t), t \geq 0\}$ of class (C_0) and if $B \in \mathscr{E}(\mathscr{X})$, then the semigroup $\{S(t), t \geq 0\}$ generated by $A + B$ with $\mathscr{D}(A + B) = \mathscr{D}(A)$ is also of class (C_0), and can be represented by the series expansion

$$S(t) = \sum_{n=0}^{\infty} S_n(t)$$

where

$$S_0(t) = T(t)$$

$$S_n(t)x = \int_0^t T(t - \tau)BS_{n-1}(\tau)x \, d\tau \qquad x \in \mathscr{D}(A)$$

Phillips[26] has considered the following problem: Given a one-parameter family of closed linear operators $A(t)$ with domains dense in a Banach space $\mathscr{X}$, find a one-parameter family of bounded linear operators $\{T(t), t \geq 0\}$ strongly continuous for $t \to 0$ such that

$$\frac{dT(t)x}{dt} = A(t)T(t)x \qquad T(0) = I$$

for all x in a given dense domain. Phillips did not give a solution to the general problem posed above, but gave a solution in the case where $A(t) = A + B(t)$; that is, he considered differential equations of the form

$$\frac{dT(t)x}{dt} = [A + B(t)] \, T(t)x$$

Because of the importance of differential equations of the above form in mathematical physics, we state the following result of Phillips:

Theorem 9: Let A be the infinitesimal generator of a semi-group $\{U(t), t \geqq 0\}$ of class (C_0). Let $B(t)$ be a strongly continuously differentiable function on $[0, \infty]$ to $\mathscr{E}(\mathscr{X})$. Then there exists a unique one-parameter family of bounded linear operators $\{T(t), t \geqq 0\}$, strongly continuous on $[0, \infty]$, such that $T(0) = I$, and for $x \in \mathscr{D}(A)$, $T(t)x$ is strongly continuously differentiable, and

$$\frac{dT(t)x}{dt} = [A + B(t)]T(t)x$$

The solution of the above differential equation can be represented as

$$T(t) = \sum_{n=0}^{\infty} S_n(t)$$

where

$$S_0(t) = U(t)$$

$$S_n(t)x = \int_0^t U(t - \tau)B(\tau)S_{n-1}(\tau)x \, d\tau$$

Returning now to Hilbert space, we give a representation theorem for contraction operators in Hilbert space due to Sz.-Nagy.[35] This theorem can be stated as follows:

Theorem 10: Let $\{T(t), t \geqq 0\}$ be a strongly continuous contraction semigroup in a Hilbert space H_0. Then there exists a group of unitary operators $\{U(t), t \in (-\infty, \infty)\}$ on a larger Hilbert space H containing H_0 as a subspace such that

$$T(t)x = PU(t)x \qquad x \in H, t \geqq 0$$

where P is a projection operator with range H_0. The space H (called the dilation space) can be constructed in a minimal fashion so that it is spanned by $\{U(t)x, x \in H_0, t \in (-\infty, \infty)\}$. In this case the structure $\{H_0, U(t), H\}$ is determined to within an isomorphism.

By Stone's theorem[33]

$$U(t) = \int_{-\infty}^{\infty} e^{i\omega t} \, dE(\omega)$$

where $E(\omega)$ is a uniquely determined resolution of the identity. Hence the contraction operator $T(t)$ admits the representation

$$T(t)x = P \int_0^\infty e^{i\omega t} \, dE(\omega)x$$

From theorem 6 we know that an operator A on Hilbert space is the infinitesimal generator of a contraction semigroup if A is maximally dissipative with dense domain. The theorem we now state, due to Dolph,[4] gives a representation of the resolvent of a maximal dissipative operator on Hilbert space:

Theorem 11: Let A be a maximally dissipative and closed operator in a Hilbert space H_0. Then for any λ, with $\mathrm{Re}(\lambda) > 0$, the resolvent operator $R(\lambda; A)$ exists and can be represented as

$$R(\lambda; A) = \int_{-\infty}^\infty \frac{dF(\omega)}{i\omega - \lambda}$$

where $F(\omega)$ [$= PE(\omega)$] is a generalized resolution of the identity, and hence the projection of an orthogonal resolution of the identity in a Hilbert space H containing H_0 as a subspace.

In view of Dolph's theorem, if $R(\lambda; A)$ is known, inversion enables us to determine $F(\omega)$, and hence $E(\omega)$, which occurs in the integral representation of the associated contraction semigroup.

As a final topic, we introduce the notion of equivalent semigroups. This notion is essentially a semigroup operator-theoretic analog of the notion of similarity in matrix theory. Let $\{T(t), t \geq 0\}$ be a semigroup of class (C_0) in a Banach space $\mathscr{Y}$. If H is a linear homeomorphism of $\mathscr{Y}$ into another Banach space $\mathscr{X}$, then the semigroup $\{S(t), t \geq 0\}$, where $S(t) = HT(t)H^{-1}$, is a semigroup of class (C_0) in $\mathscr{X}$. The two semigroups are said to be homeomorphically equivalent or similar.

Definition 9. Two semigroups, $\{T(t), t \geq 0\}$ and $\{S(t), t \geq 0\}$ say, are said to be equivalent if there exist real constants ω and α, with α positive, such that $S(t)$ and $e^{\omega t} T(\alpha t)$ are homeomorphically equivalent:

$$S(t) = H[e^{\omega t} T(\alpha t)]H^{-1}$$

If A and B are the infinitesimal generators of the semigroups $\{T(t), t \geq 0\}$ and $\{S(t), t \geq 0\}$, respectively, then the following relations can be established:

(A) $B = \omega I + \alpha H A H^{-1}$
(B) $\mathscr{D}(B) = H\mathscr{D}(A)$
(C) $R(\lambda; B) = HR(\lambda - \omega, \alpha A)H^{-1}$

Equivalent semigroups have found applications in partial differential equations[23] and Markov processes,[1] and may be of interest in quantum mechanics (see Section 4D).

3. SEMIGROUPS ASSOCIATED WITH SOME DIFFERENTIAL EQUATIONS OF MATHEMATICAL PHYSICS

A. Introduction

In this section, we shall study the semigroups of operators associated with two of the well-known differential equations of mathematical physics. In Section B, we study the equation of diffusion and heat conduction, and in Section C we study the equation of the vibrating string. For additional material on the semigroups, considered in these sections we refer to Hille.[11]

For other studies utilizing semigroup theory to study differential equations of mathematical physics, we refer to the papers of Feller,[8] Kato,[16] and Yosida.[37-40]

B. The Equation of Diffusion and Heat Conduction

We will consider the problem of diffusion or heat conduction on the real line. As is well known, the equation in this case is the parabolic equation

$$\frac{\partial u}{\partial t} = \frac{\partial^2 u}{\partial x^2} \qquad u = u(t, x)$$

$$u(t, 0) = f(x) \tag{3.1}$$

The solution of equation (3.1) is

$$u(t, x) = \int_{-\infty}^{\infty} K(x - \xi, t) f(\xi) \, d\xi \tag{3.2}$$

where

$$K(x, t) = (4\pi t)^{-1/2} e^{-(x^2/4t)} \tag{3.3}$$

We can write (3.2) in the form

$$u(t, x) = T(t) f(x) \tag{3.4}$$

where $T(t)$ is an integral operator with kernel given by (3.3). It is easily verified that $T(t)$ is a semigroup operator, and therefore $\{T(t), t \geqq 0\}$ is a semigroup of operators. We now investigate this semigroup.

In this case, we can take as the Banach space $\mathscr{X}$ the space of continuous functions $C(-\infty, \infty)$ or the Lebesgue space $L_p(-\infty, \infty)$, where p is fixed, $1 \leq p < \infty$. In either case, $\{T(t),\ t \geq 0\}$ is a strongly continuous semigroup of operators in $\mathscr{X}$. Further, the initial condition is satisfied in the sense that $\lim_{t \to 0} \|u(t,\ x) - f(x)\| = 0$ for every $f(x) \in \mathscr{X}$. We can easily show that $\|T(t)\| \leq 1$, so that $\{T(t),\ t \geq 0\}$ is a contraction semigroup of class (C_0).

Let us now consider the infinitesimal generator of $\{T(t),\ t \geq 0\}$. Equation (3.1) can be written in the form

$$\frac{\partial u}{\partial t} = Au$$

where

$$A \cdot = \frac{\partial^2 \cdot}{\partial x^2} \tag{3.5}$$

We now state the following result, which proves that (3.5) is the infinitesimal generator of the semigroup $\{T(t),\ t \geq 0\}$:

Theorem 12: If $f(x)$, $f'(x)$, and $f''(x)$ are elements of $\mathscr{X}$, in either of the two cases considered above, then

$$\lim_{t \to 0} \left\| \frac{[T(t) - I]f(x)}{t} - f''(x) \right\| = 0$$

We close this section by considering a semigroup of operators equivalent to $\{T(t),\ t \geq 0\}$. For $f(x) \in L_2\,(-\infty,\ \infty)$, let

$$f(\lambda) = \lim_{n \to \infty} \frac{1}{\sqrt{2\pi}} \int_{-n}^{n} e^{i\lambda x} f(x)\,dx$$

be the Fourier–Plancherel transform of $f(x)$. In terms of $f(\lambda)$ we can write

$$u(t,\ x) = \frac{1}{\sqrt{2\pi}} \int_{-\infty}^{\infty} e^{-i\lambda x - \lambda^2 t} f(\lambda)\,d\lambda \tag{3.6}$$

Let $\mathscr{Y}$ be the space of all such Fourier–Plancherel transforms of elements of $L_2(-\infty,\ \infty)$. We know that the mapping $H \colon L_2(-\infty,\ \infty) \to \mathscr{Y}$ defined by

$$H[f(x)] = f(\lambda) \tag{3.7}$$

is an isometric (and thus homeomorphic) isomorphism. The inverse mapping $H^{-1} \colon \mathscr{Y} \to L_2(-\infty,\ \infty)$ is, of course, defined by the inverse transformation

$$H^{-1}f(\lambda) = f(x) = \lim_{n \to \infty} \frac{1}{\sqrt{2\pi}} \int_{-n}^{n} e^{-i\lambda x} f(\lambda)\,d\lambda \tag{3.8}$$

On the space $\mathscr{Y}$ let $\{S(t), t \geq 0\}$ be the semigroup defined by

$$S(t)[f(\lambda)] = e^{-\lambda^2 t} f(\lambda) \qquad (3.9)$$

Now, for $f(x) \in L_2(-\infty, \infty)$

$$
\begin{aligned}
[H^{-1}S(t)H]f(x) &= [H^{-1}S(t)]H[f(x)] \\
&= [H^{-1}S(t)]f(\lambda) \\
&= H^{-1}[e^{-\lambda^2 t}f(\lambda)] \\
&= \frac{1}{\sqrt{2\pi}} \int_{-\infty}^{\infty} e^{-i\lambda x} e^{-\lambda^2 t} f(\lambda)\, d\lambda \\
&= \frac{1}{\sqrt{2\pi}} \int_{-\infty}^{\infty} e^{-i\lambda x - \lambda^2 t} f(\lambda)\, d\lambda \\
&= u(t, x) \quad \text{by (3.6)} \\
&= T(t)f(x) \quad \text{by (3.4)}
\end{aligned}
$$

Thus

$$T(t) = H^{-1}S(t)H \qquad (3.10)$$

or

$$S(t) = HT(t)H^{-1} \qquad (3.11)$$

Therefore, we have shown that the semigroups $\{T(t), t \geq 0\}$ and $\{S(t), t \geq 0\}$ are equivalent with $\omega = 0$, $\alpha = 1$, and H as defined by (3.7).

C. The Equation of the Vibrating String

We consider the initial-value problem given by the hyperbolic equation

$$\frac{\partial^2 u}{\partial t^2} = \frac{\partial^2 u}{\partial x^2} \qquad u = u(t, x)$$

$$u(0, x) = f_1(x) \qquad (3.12)$$

$$\left. \frac{\partial u}{\partial t} \right|_{t=0} = f_2(x)$$

The solution of equation (3.12) is

$$u(t, x) = \tfrac{1}{2}\{f_1(x + t) + f_1(x - t)\} + \tfrac{1}{2}\int_{x-t}^{x+t} f_2(\xi)\, d\xi \qquad (3.13)$$

We assume that the derivative of $f_1(x)$ is absolutely continuous, and that $f_2(x)$ is absolutely continuous. Then $u(t, x)$ as given by (3.13) satisfies (3.12) for almost all (t, x).

In this case, let $\mathscr{X}$ be the space of vector functions $F(x) = [f_1(x), f_2(x)]$, where $f_1(x)$ and $f_2(x)$ are bounded and absolutely continuous together with their first derivatives on $(-\infty, \infty)$. We define the norm $\|F\|$ as the supremum of the absolute values of the functions $f_1(x)$, $f_2(x), f_1'(x)$, and $f_2'(x)$. With the norm so defined, $\mathscr{X}$ is a Banach space. If we put $F(t, x) = [u(t, x), \partial u/\partial t]$, we can write

$$F(t, x) = T(t)[F(x)] \tag{3.14}$$

where

$$T(t) = \begin{pmatrix} g'(t) & g(t) \\ g''(t) & g'(t) \end{pmatrix} \tag{3.15}$$

with

$$g(t)[f] = \tfrac{1}{2} \int_{x-t}^{x+t} f(\xi)\, d\xi \tag{3.16}$$

Clearly, $T(0) = I$, and a rather involved calculation shows that $T^{-1}(t)$ exists, and $T(s + t) = T(s)\, T(t)$; so that $\{T(t), t \in (-\infty, \infty)\}$ is a group of operators.

Since the operator $T(t)$ is a matrix, its infinitesimal generator will also be a matrix, but the identification of this operator with the right-hand side of equation (3.12) will be obvious. We have, by definition

$$A_\xi = \frac{T(\xi) - I}{\xi} = \begin{pmatrix} \dfrac{g'(\xi) - 1}{\xi} & \dfrac{g(\xi)}{\xi} \\ \dfrac{g''(\xi)}{\xi} & \dfrac{g'(\xi) - 1}{\xi} \end{pmatrix} \tag{3.17}$$

Now, if $f'(x)$ is continuous, the diagonal elements

$$\frac{[g'(\xi) - I]f(x)}{\xi} = \frac{1}{2\xi} \{f(x + \xi) + f(x - \xi) - 2f(x)\}$$

$$= \frac{1}{2\xi} \{f'(x + \theta_1 \xi) - f'(x - \theta_2 \xi)\} \to 0$$

as $\xi \to 0$. Also, if $f(x)$ is continuous

$$\frac{g(\xi)[f(x)]}{\xi} = \frac{1}{2\xi} \int_{x-\xi}^{x+\xi} f(\xi)\, d\xi \to f(x)$$

as $\xi \to 0$.

Finally,

$$\frac{g''(\xi)[f(x)]}{\xi} = \frac{1}{2\xi} [f'(x + \xi) - f'(x - \xi)] \to f''(x)$$

whenever the limit exists. Hence, it follows that

$$A = \lim_{\xi \to 0} A_\xi = \begin{pmatrix} 0 & I \\ \dfrac{\partial^2}{\partial x^2} & 0 \end{pmatrix} \tag{3.18}$$

is the infinitesimal generator of the group $\{T(t), t \in (-\infty, \infty)\}$.

4. SEMIGROUPS OF OPERATORS IN QUANTUM MECHANICS

A. Introduction

In this section, we consider some applications of semigroup methods in quantum mechanics. Section B is devoted to a brief discussion of the Schrödinger equation of quantum mechanics from the point of view of semigroup theory. We also give as a simple example the semigroup associated with the classical harmonic oscillator. In Section C we study semigroup and groups of dynamic mappings in quantum mechanics, and in Section D we make a few remarks about equivalent quantum mechanical system.

B. The Schrödinger Equation from the Point of View of Semigroup Theory

In the Schrödinger representation, the wave function satisfies the differential equation

$$\frac{\partial \psi}{\partial t} = -\frac{iH}{\hbar} \psi \tag{4.1}$$

where H is the Hamiltonian operator, and, in the one-dimensional case, $\psi = \psi(t, x)$. Equation (4.1) is the abstract form of Schrödinger's equation, and can be regarded as a differential equation in a concrete Hilbert space $\mathscr{H}$. The solution of equation (4.1) can be expressed in the form

$$\psi(t, x) = \exp\left[(-iH/\hbar)t\right]\psi(0, x) \tag{4.2}$$

where $\psi(0, x)$ is some suitably chosen function in $\mathscr{H}$. If we introduce the operator-valued function $S(t) = \exp\left[(-iH/\hbar)t\right]$, then equation (4.2) can be written as

$$\psi(t, x) = S(t)\left[\psi(0, x)\right] \tag{4.3}$$

Because of the exponential character of $S(t)$, it is clear that $S(0) = I$ and that for all $s, t > 0$ $S(s + t) = S(s) S(t)$, that is, $\{S(t), t \geqq 0\}$ is a

semigroup of operators in $\mathcal{H}$. We will call this semigroup the Schrödinger semigroup.

In the case where the Hilbert space is L_2, a semigroup can be
defined for a dense set of functions as

$$T(t)\,[g(x)] = \int K(t, x, dy)g(y)$$

where K is the unitary quantum mechanical kernel function.

The problem now is that of determining when a given Hamiltonian
operator is the infinitesimal generator of a Schrödinger semigroup.
This problem can be resolved by the theorems given in Section 2C.
We recall that an operator H on a Hilbert space $\mathcal{H}$ is said to be conservative if $(Hx, x) + (x, Hx) = 0$, and that H is conservative if and
only if iH is symmetric. Further, if H is conservative and maximally
dissipative with dense domain, then iH is maximally symmetric.
Theorem 7 then states that a necessary and sufficient condition that
a Hamiltonian H generate a Schrödinger semigroup of isometric
operators is that iH be maximally symmetric (i.e., H be conservative
and maximally dissipative with dense domain).

Since $S(t)$ is an isometric operator, we can use Sz.-Nagy's representation theorem to obtain the representation

$$S(t) = PU(t) \qquad (4.4)$$

It $\mathcal{H}$ is already its own dilation space, then $P = I$, and the Schrödinger
semigroup is given by a semigroup of unitary operators. From Stone's
theorem, we obtain

$$S(t) = \int_{-\infty}^{\infty} e^{i\lambda t}\,dE(\lambda) \qquad (4.5)$$

In order to obtain $S(t)$ explicitly, the resolution of the identity $\{E(\lambda)\}$
must be determined. At this stage, Dolph's theorem, together with an
inversion operation, can be used to determine $E(\lambda)$.

We close this section by considering the semigroup of operators
associated with the classical harmonic oscillator. In classical mechanics,
the harmonic oscillator of unit mass and frequency ν is described by
the Hamilton equations

$$\frac{dp}{dt} = -\alpha^2 q$$

$$\frac{dq}{dt} = p \qquad\qquad (4.6)$$

where q and p are the position coordinate and momentum, respectively, and $\alpha^2 = 2\pi v^2$. The solution of equation (4.6) is

$$q(t) = q(0) \cos \alpha t + \tfrac{1}{2} p(0) \sin \alpha t$$
$$p(t) = -q(0) \alpha \sin \alpha t + p(0) \cos \alpha t \tag{4.7}$$

which can be written in matrix form as

$$\begin{pmatrix} q(t) \\ p(t) \end{pmatrix} = \begin{pmatrix} \cos \alpha t & \dfrac{1}{\alpha} \sin \alpha t \\ -\alpha \sin \alpha t & \cos \alpha t \end{pmatrix} \begin{pmatrix} q(0) \\ p(0) \end{pmatrix} \tag{4.8}$$

Let us put

$$T(t) = \begin{pmatrix} \cos \alpha t & \dfrac{1}{\alpha} \sin \alpha t \\ -\alpha \sin \alpha t & \cos \alpha t \end{pmatrix} \tag{4.9}$$

We will show that $\{T(t),\ t \geq 0\}$ is a semigroup of operators. Clearly $T(0) = I$. For $t_1, t_2 > 0$, we have

$$q(t_1 + t_2) = q(C) \cos \alpha(t_1 + t_2) + \frac{p(0)}{\alpha} \sin \alpha(t_1 + t_2)$$

$$p(t_1 + t_2) = p(0) \cos \alpha(t_1 + t_2) - \alpha q(0) \sin \alpha(t_1 + t_2)$$

which can be rewritten as

$$q(t_1 + t_2) = \left[q(0) \cos \alpha t_2 + \frac{p(0)}{\alpha} \sin \alpha t_2 \right] \cos \alpha t_1$$
$$+ \left[\frac{p(0) \cos \alpha t_2 - \alpha q(0) \sin \alpha t_2}{\alpha} \right] \sin \alpha t_1$$

$$p(t_1 + t_2) = [p(0) \cos \alpha t_2 - \alpha q(0) \sin \alpha t_2] \cos \alpha t_1$$
$$- \alpha \left[q(0) \cos \alpha t_2 + \frac{p(0)}{\alpha} \sin \alpha t_2 \right] \sin \alpha t_1$$

Hence, from (4.9) and the above calculations, we obtain

$$T(t_1 + t_2) = T(t_1) T(t_2), \quad t_1, t_2 > 0$$

Let us now consider the infinitesimal generator A of $\{T(t),\ t \geq 0\}$. From equations (4.6) and (4.9), it is clear that

$$A = \begin{pmatrix} 0 & 1 \\ -\alpha^2 & 0 \end{pmatrix} \tag{4.10}$$

Hence, equation (4.6) could have been written as

$$\frac{d}{dt} \begin{pmatrix} q(t) \\ p(t) \end{pmatrix} = \begin{pmatrix} 0 & 1 \\ -\alpha^2 & 0 \end{pmatrix} \begin{pmatrix} q(t) \\ p(t) \end{pmatrix} \tag{4.11}$$

For the quantum harmonic oscillator, the Schrödinger equation is given by

$$\frac{\partial \psi}{\partial t} = \frac{\hbar^2}{2} \cdot \frac{\partial^2 \psi}{\partial x^2} + \tfrac{1}{2} \alpha^2 x^2 \psi \tag{4.12}$$

and this equation can be studied using the matrix approach used in the study of the vibrating string (Section 3C).

C. Semigroups of Dynamical Mappings of Density Operators

It is well known that the quantum mechanical state of a physical system can be specified by a density operator D which satisfies the following conditions[6, 25]:

(A) $(\phi, D\psi) = (D\phi, \psi)$ (Hermiticity)
(B) $(\phi, D\phi) \geq 0$ (Positive-definiteness)
(C) $\mathrm{Tr}(D) = 1$ (Normalization)

where ϕ and ψ are any elements of the Hilbert space on which the operator D is defined. The density operators form a convex set, the extremal elements of which are the pure-state density operators, and these elements are projection operators onto one-dimensional subspaces of $\mathscr{H}$. It has been shown that the density operators belong to the Hilbert space $\mathscr{D}$ of operators D on $\mathscr{H}$ for which $\mathrm{Tr}(D^* D)$ is finite, the inner product in $\mathscr{D}$ being defined by $(D_1, D_2) = \mathrm{Tr}(D_1^* D_2)$.

In a recent series of papers, Jordan and Sudarshan[14, 15, 34] have studied dynamical mappings of density operators. Let T be a linear operator on $\mathscr{D}$ such that, if D is a density operator, then

$$D' = TD \tag{4.13}$$

is also a density operator. Such an operator T is called a dynamical mapping. It is clear that in order to represent dynamics in the usual sense, that is, as a continuous time-dependent evolution of the state of a physical system, we must introduce, in place of (4.13), the mapping

$$D \to D(t) = T(t)D \tag{4.14}$$

where the dynamical mapping T is a function of the real parameter t. From stationarity considerations, we require that $T(s + t) = T(s) T(t)$, and that $T(0) = I$; that is to say, we require that the family of dynamical mappings $\{T(t), t \geq 0\}$ form a one-parameter semigroup. Let F be a self-adjoint operator belonging to $\mathscr{D}$. We also require that the expectation value

$$E_t\{F\} = \mathrm{Tr}\big(F, D(t)\big) = \big(F, T(t)D\big) \qquad (4.15)$$

be a continuous function of t. Since the trace of the product is expressed as the inner product in $\mathscr{D}$, the above requirement is equivalent to requiring that the semigroup operator $T(t)$ be weakly continuous as a function of t.

Let us now consider various "models" for the evolution of density operators.

1. The mathematical model for the evolution of density operators as described above is as follows:

(A) $\{T(t), t \geqq 0\}$ is a weakly continuous semigroup of dynamical mappings in $\mathscr{D}$.

2. If the dynamics is to be reversible, that is, we require every dynamical mapping $T(t)$ to have an inverse $T^{-1}(t) = T(-t)$, then the mathematical model becomes:

(B) $\{T(t), t \in (-\infty, \infty)\}$ is a weakly continuous group of dynamical mappings in $\mathscr{D}$.

3. A dynamical mapping is said to be a Hamiltonian dynamical mapping if it admits a representation as a unitary operator. In order to represent Hamiltonian dynamics we require the following model:

(C) There exists a (strongly or weakly) continuous group of unitary operators $\{U(t), t \in (-\infty, \infty)\}$ on $\mathscr{D}$ such that $T(t)\,D = U(t)\,DU^*(t)$ for each $D \in \mathscr{D}$.

The relationship between the above models can be expressed as follows:

$$C \Longrightarrow B \Longrightarrow A$$
$$B \longrightarrow C \text{ and } A \not\longrightarrow B$$

The main theorem given in Jordan and Sudarshan[15] can be stated as follows:

Theorem 13: A sufficient condition that a family of dynamical mappings $T(t)$ represent Hamiltonian dynamics is that $\{T(t), t \in (-\infty, \infty)\}$ be a weakly continuous group.

We close this section by remarking that there are several open problems which seem to be of interest. Some of these are as follows:

(1) What operator A is the infinitesimal generator of the group of operators, and how is A related to the Hamiltonian operator in the Schrödinger representation? We refer to equation (4.16) for a differential equation relationship between $D(t)$ and H.

(2) What generation theorems are required in the study of dynamical

mappings, and how do the conditions of these theorems reflect the "physical aspects" of the problem?

(3) What, if any, is the relationship between the Schrödinger semi-group $\{S(t), t \geqq 0\}$ and the group of dynamical mapping? It is known, for example, that if the Hamiltonian is self-adjoint, then the differential equation for $D(t)$ is given by

$$\frac{\partial D}{\partial t} = -\frac{1}{\hbar}(HD - DH) \tag{4.16}$$

and in this case

$$D(t) = e^{-\frac{iH}{\hbar}t} D(0) e^{\frac{iH}{\hbar}t}$$
$$= S(t)D(0)S^*(t) \in \mathscr{D} \tag{4.17}$$

D. Equivalent Quantum-Mechanical Systems

In Section 2D, we introduced the notion of an equivalent semigroup of operators. In this section, we consider what will be called "equivalent quantum-mechanical systems" and make a few brief remarks about such systems.

We will define a quantum-mechanical system to be a triple $\{\mathscr{H}, S(t), H\}$ where $\mathscr{H}$ is a concrete Hilbert space, $S(t)$ is a Schrödinger semigroup operator, and H is the Hamiltonian. Let $\{\mathscr{H}_1, S_1(t), H_1\}$ and $\{\mathscr{H}_2, S_2(t), H_2\}$ be two quantum-mechanical systems. We will say that the two systems are equivalent if the Schrödinger operators are equivalent in the sense of Definition 9, that is

$$S_2(t) = M((e^{\omega t} S_1(\alpha t)) M^{-1} \tag{4.18}$$

where M is a linear homeomorphism of $\mathscr{H}_1$ onto $\mathscr{H}_2$, and ω and α are real constants, with α positive.

An example of equivalent quantum-mechanical systems is given by the well-known equivalence of the Heisenberg and Schrödinger representations.[13, 21, 25] Let ψ_H and ψ_S represent the wave functions in the Heisenberg and Schrödinger representations, respectively. Their equivalence, with $\omega = 0$ and $\alpha = 1$ follows from the following relations:

$$\psi_S = S(t)[\psi(0, x)] = e^{-\frac{iH}{\hbar}t}[\psi(0, x)]$$
$$\psi_H = \psi(0, x)$$
$$\psi_S = S(t)[\psi_H]$$
$$\psi_H = S^{-1}(t)[\psi_{(S)}]$$

The question of interest to us is how this notion of equivalence might be used in the solution of operator problems in quantum mechanics. In particular, we are interested in using the notion of equivalence to study the properties of a complex quantum-mechanical system by utilizing the properties of an equivalent known system which is not so complex.

The relationship between quantum-mechanical processes and Markov processes has been discussed by several authors.[22, 31] We have discussed this relationship from the point of view of equivalent semigroups.[3]

5. SEMIGROUPS AS SOLUTIONS OF THE BOLTZMANN EQUATION OF TRANSPORT THEORY

A. Introduction

In this section, we show how semigroup theory can be used to obtain the solution of the linearized Boltzmann equation of transport theory. We will consider the case of an infinite plane slab of transport material extending from $-a \leqq z \leqq a$. Let $\sigma\Delta + 0(\Delta)$ denote the probability of a collision occurring between a fixed nucleus and a particle moving between z and $z + \Delta(\Delta > 0)$ in either direction; σ is called the cross section, and we assume that it is constant. We also assume that the production of particles is isotropic, i.e., the direction of particles arising from a collision is independent of the colliding particle. The slab is assumed to be surrounded by a vacuum, so that no particles may enter the slab from the outside.

The problem may be formulated as follows:

$$\frac{1}{c}\frac{\partial N}{\partial t} + \mu\frac{\partial N}{\partial z} + \sigma N = \frac{\gamma}{2}\sigma \int_{-1}^{1} N(z, \xi, t)\, d\xi \qquad (5.1)$$

where $N = N(z, \mu, t)$ denotes the density of the neutron beam in directions with z-direction cosine μ, and t denotes time. We have the boundary and initial conditions

$$\left.\begin{array}{ll} N(a, \mu, t) = 0 & \mu < 0, t > 0 \\ N(-a, \mu, t) = 0 & \mu > 0, t > 0 \end{array}\right\} \qquad (5.2)$$

$$N(z, \mu, 0) = f(z, \mu) \quad -a \leqq z \leqq a \qquad (5.3)$$

$$-1 \leqq \mu \leqq 1$$

Equation (5.1) can be simplified by putting

$$n(z, \mu, t) = e^{\sigma c t} N(z, \mu, t)$$

and then putting $c = 1$ and $\sigma = 1$. Then equation (5.1) becomes

$$\frac{\partial n}{\partial t} = -\mu \frac{\partial n}{\partial t} + \frac{\gamma}{2} \int_{-1}^{1} n(z, \xi, t) \, d\xi \qquad (5.4)$$

The boundary and initial conditions are not changed, but in the new notation we write

$$\left. \begin{array}{ll} n(a, \mu, t) = 0 & \mu < 0, t > 0 \\ n(-a, \mu, t) = 0 & \mu > 0, t > 0 \end{array} \right\} \qquad (5.5)$$

$$n(z, \mu, 0) = f(z, \mu) \quad -a \leq z \leq a \qquad (5.6)$$

$$-1 \leq \mu \leq 1$$

B. Solution of the Boltzmann Equation

Let us now write equation (5.4) as the operator equation

$$\frac{\partial n}{\partial t} = An \qquad (5.7)$$

where

$$A \cdot = -\mu \frac{\partial \cdot}{\partial z} + \frac{\gamma}{2} \int_{-1}^{1} \cdot \, d\xi \qquad (5.8)$$

In order to consider the operator equation (5.7) from the point of view of semigroup theory, the first thing to do is select a concrete Banach space, the elements of which are the functions $n(z, \mu, t)$, and on which A operates. As our space we choose the Hilbert space $\mathcal{H}$ of complex valued functions $g(z, \mu)$ defined and Lebesgue-square-integrable over the rectangle $|z| \leq a, \mu \leq 1$, that is

$$\int_{-1}^{1} \int_{-a}^{a} |g(z, \mu)|^2 \, dz \, d\mu < \infty$$

For functions g, h in $\mathcal{H}$, the inner product, as usual, is given by

$$(g, h) = \int_{-1}^{1} \int_{-a}^{a} g(z, \mu) \overline{h(z, \mu)} dz \, d\mu$$

and the norm of an element $g \in \mathcal{H}$ is given by

$$\|g\| = \sqrt{(g, g)} = \left\{ \int_{-1}^{1} \int_{-a}^{a} |g(z, \mu)|^2 \, dz \, d\mu \right\}^{1/2}$$

Hence

$$\mathcal{H} = \{g(z, \mu): |z| \leq a, |\mu| \leq 1, \|g\| < \infty\}$$

Let us now rewrite the operator A as

$$A = -D + \gamma J \qquad (5.9)$$

where

$$D \cdot = \mu \frac{\partial \cdot}{\partial z} \qquad (5.10)$$

and

$$J \cdot = \tfrac{1}{2} \int_{-1}^{1} \cdot \, d\xi \qquad (5.11)$$

Hence A can be regarded as the operator obtained by the perturbation of the operator $-D$ by the operator γJ. Let $\mathcal{D}(D)$ be the set of functions $g \in \mathcal{H}$ such that g is absolutely continuous in z for each μ such that $|\mu| \leq 1$, and such that $Dg \in \mathcal{H}$; and let $\mathcal{D}(J)$ be the set of functions $g \in \mathcal{H}$ such that Jg exists for each z such that $|z| \leq a$, and $Jg \in \mathcal{H}$. Finally, $\mathcal{D}(A)$ is defined as the set of functions $g \in \mathcal{D}(D) \cap \mathcal{D}(J)$, and such that

$$\left.\begin{array}{ll} g(a, \mu) = 0 & -1 \leq \mu < 0 \\ g(-a, \mu) = 0 & 0 < \mu \leq 1 \end{array}\right\} \qquad (5.12)$$

Hence A is a linear operator on $\mathcal{H}$ with domain $\mathcal{D}(A)$. We remark that the operator adjoint to A is

$$A \cdot {}^{*} = \mu \frac{\partial \cdot}{\partial z} + \frac{\gamma}{2} \int_{-1}^{1} \cdot \, d\xi \qquad (5.13)$$

Thus, unfortunately, A is not a self-adjoint operator.

As an initial-value problem, we seek a solution of equation (5.7) of the form

$$n(z, \mu, t) = T(t)[f(z, \mu)] \qquad (5.14)$$

where $\{T(t), t \geq 0\}$ is a semigroup of operators in $\mathcal{H}$. In order to obtain such a solution there are two things that need to be done: (a) we must characterize the spectrum and resolvent set of A, and (b) we must show that A is the infinitesimal generator of a semigroup of operators in $\mathcal{H}$. We refrain from a discussion of (a) and (b), since it is rather involved and lengthy and the interested reader can find this discussed elsewhere.[17, 18, 36] We simply give a summary of the known results:

Theorem 14: The spectrum and resolvent set of the linear operator A defined by (5.8) are as follows:

$P\sigma(A) = $ a finite non-empty point-set lying on $\lambda > 0$.

$R\sigma(A) = $ empty set.

$C\sigma(A) = \{\lambda: \mathrm{Re}(\lambda) \leq 0\}$.

$\rho(A) = \{\lambda: \mathrm{Re}(\lambda) > 0$, deleted by $P\sigma(A)\}$.

We will denote by $\beta_1 > \beta_2 > \ldots > \beta_k, k \geq 1$ the elements of the point spectrum, and denote by

$$\psi_1(z, \mu) \ \psi_2(z, \mu), \ldots, \psi_k(z, \mu)$$

the associated eigenfunctions.

To turn now to the generation problem, we first observe that since A is a linear operator on a Hilbert space, there are two methods available for us to use in determining if A is the infinitesimal generator of a semigroup: First, we could employ the Hille–Yosida theorem (theorem 5), or, second, we could show that A is a maximally dissipative operator and use Phillips'[8,36] theorem. The Hille–Yosida theorem has been used to show that A is indeed the infinitesimal generator of a semigroup; however, it can be shown that A is maximally dissipative with dense domain, and hence Phillips' theorem is applicable. Therefore, we can conclude that A is the infinitesimal generator of a strongly continuous semigroup of operators of class (C_0) in $\mathscr{H}$, say $\{T(t), t \geq 0\}$. Moreover, by Theorem 1, $T(t)$ is strongly differentiable, and for $f \in \mathscr{D}(A)$

$$\frac{dT(t)f}{dt} = AT(t)f = T(t)Af \tag{5.15}$$

If we now put

$$n(z, \mu, t) = T(t)f(z, \mu) \qquad f \in D(A)$$

we see that equation (5.7) is satisfied, and that the initial condition is satisfied in the form

$$\lim_{t \to 0} \|n(z, \mu, t) - f(z, \mu)\| = 0$$

and the boundary conditions are satisfied, because from equation (5.15) $n = T(t)f \in \mathscr{D}(A)$. Finally, the uniqueness of the solution follows from Theorem 9 with $B(t) = 0$, the null operator.

Since the semigroup is of class (C_0), by Theorem 3 the solution $n(z, \mu, t)$ can be represented as a Laplace integral, that is

$$n(z, \mu, t) = \lim_{\omega \to \infty} \frac{1}{2\pi i} \int_{\alpha - i\omega}^{\alpha + i\omega} e^{\lambda t} R(\lambda, A)f \, d\lambda \tag{5.16}$$

for $t > 0$, $\alpha > \beta$, $f \in \mathscr{D}(A)$. We summarize the results by stating that the solution of equation (5.7) is given by

$$n(z, \mu, t) = \sum_{i=1}^{k} (f, \psi_i^*)\psi_i(z, \mu)e^{\beta_i t} + \xi(z, \mu, t) \qquad (5.17)$$

where

$$\xi(\partial, \mu, t) = \lim_{\omega \to \infty} \frac{1}{2\pi i} \int_{\alpha - i\omega}^{\alpha + i\omega} e^{\lambda t} R(\lambda, A) f \, d\lambda \quad 0 < \alpha < \beta_\kappa$$

and $(f, \psi_i^*)\psi_i$ is the residue of the pole of $R(\lambda; A)f$ at $\lambda = \beta_i$, and ψ_i^* is the adjoint eigenfunction.

C. Other Studies on the Boltzmann Equation

We refer to Wing[36] for a summary of recent studies using semigroup methods in neutron transport theory.

REFERENCES

1. A.T. Bharucha-Reid, "Processus de Markov équivalents et semigroups d'opérateurs," *Compt. Rend. Acad. Sci. Paris* **257**, 1668 (1963).
2. A.T. Bharucha-Reid, "Lectures on Semigroups of Operators," Institute of Mathematical Sciences, Madras, India (1964).
3. A.T. Bharucha-Reid, "Note on Markov Processes and Quantum Mechanical Processes." (To be published.)
4. C.L. Dolph, "Positive Real Resolvents and Linear Passive Hilbert Systems," *Ann. Acad. Sci. Fenn.*, Ser. A I, No. 336 (1963).
5. N. Dunford and J.T. Schwartz, *Linear Operators. Part I. General Theory*, Interscience Publishers, Inc., New York (1958).
6. U. Fano, "Description of States in Quantum Mechanics by Density Matrix and Operator Techniques," *Rev. Mod. Phys.* **29**, 74 (1957).
7. J. Feldman, "On the Schrödinger and Heat Equations for a Non-negative Potential, "*Trans. Amer. Math. Soc.* **108**, 251 (1963).
8. W. Feller, "On the Equation of the Vibrating String," *J. Math. Mech.* **8**, 339, (1959).
9. J. Hadamard, *Lectures on Cauchy's Problem in Partial Differential Equations*, Yale University Press, New Haven (1923).
10. E. Hille, "Notes on Linear Transformations. I," *Trans. Amer. Math. Soc.* **39**, 131 (1936).
11. E. Hille, *Functional Analysis and Semigroups*, American Mathematical Society, New York (1948).
12. E. Hille and R.S. Phillips, *Functional Analysis and Semi-Groups*, American Mathematical Society, New York (1957).
13. E. Ikenberry, *Quantum Mechanics*, Oxford University Press, New York (1962).

14. T.F. Jordan, M.A. Pinsky, and E.C.G. Sudarshan, "Dynamical Mappings of Density Operators in Quantum Mechanics. II," *J. Math. Phys.* 3, 848 (1962).

15. T.F. Jordan and E.C.G. Sudarshan, "Dynamical Mappings of Density Operators in Quantum Mechanics," *J. Math Phys.* 2, 772 (1961).

16. T. Kato, "On Linear Differential Equations in Banach Spaces," *Comm. Pure Appl. Math.* 9, 479 (1956).

17. J. Lehner, and G.M. Wing, "On the Spectrum of an Unsymmetric Operator Arising in the Transport Theory of Neutrons," *Comm. Pure. Appl. Math.* 8, 217 (1955).

18. J. Lehner and G.M. Wing, "Solution of the Linearized Boltzmann Equations for the Slab Geometry," *Duke Math. J.* 23, 125 (1956).

19. G.W. Mackey, *Mathematical Foundations of Quantum Mechanics*, W.A. Benjamin, Inc., New York (1963).

20. K. Maurin, *Methods of Hilbert Space*, Panstwowe Wydawictwo Naukowe, Warsaw (1959).

21. J. McConnell, *Quantum Particle Dynamics*, North Holland Publishing Co., Amsterdam (1960).

22. E.W. Montroll, "Markoff Chains, Wiener Integrals, and Quantum Mechanics, *Comm. Pure Appl. Math.* 5, 415 (1952).

23. T.W. Mulliken, "Semi-Groups of Operators of Class (C_0) in L_p Determined by Parabolic Differential Equations," *Pacific J. Math.* 9, 791 (1959).

24. E.·Nelson, "Feynman Integrals and the Schrödinger Equation," *J. Math. Phys.* 5, 332 (1964).

25. J. Von Neumann, *Mathematical Foundations of Quantum Mechanics*, Princeton University Press, Princeton (1955).

26. R.S. Phillips, "Perturbation Theory for Semi-Groups of Linear Operators," *Trans. Amer. Math. Soc.* 74, 191 (1953).

27. R.S. Phillips, "Semigroups of Operators," *Bull. Amer. Math. Soc.* 61, 16 (1955).

28. R.S. Phillips. "Dissipative Operators and Hyperbolic Systems of Partial Differential Equations," *Trans. Amer. Math. Soc.* 90, 193 (1959).

29. R.S. Phillips, "Semigroup Methods in the Theory of Partial Differential Equations," *Modern Mathematics for the Engineer*, Second Series, McGraw-Hill Book Company, New York (1961), pp. 100–132.

30. F. Riesz, and B. Sz.-Nagy, *Lecons d'analyse fonctionelle*, Akademia i Kiado, Budapest (1953).

31. H. Rubin, "On the Foundations of Quantum Mechanics," *Symposium on the Axiomatic Method*, Amsterdam (1959), North Holland Publishing Co., pp. 333–340.

32. I.E. Segal, "Non-Linear Semi-Groups," *Ann Math.* 78, 339 (1963).

33. M.H. Stone, "On One-Parameter Unitary Groups in Hilbert Space," *Ann. Math.* 33, 643 (1932).

34. E.C.G. Sudarshan, *Lectures on Foundation of Quantum Mechanics and Field Theory*. Madras: Institute of Mathematical Sciences, 1962.

35. B. Sz.-Nagy, "Sur les contractions de l'espace de Hilbert," *Acta Sci. Math. Szeged* 15, 87 (1953).

36. G.M. Wing, *An Introduction to Transport Theory*, John Wiley and Sons, New York (1962).

37. K. Yosida, "An Operator-Theoretical Integration of the Wave Equation," *J. Math. Soc. Japan* **8**, 79 (1956).
38. K. Yosida, "Integration of the Wave Equation by the Theory of Semi-Groups," *Sugaku* **8**, 65 (1956–1957).
39. K. Yosida, "An Operator-Theoretical Integration of the Temporally Inhomogeneous Wave Equation," *J. Fac. Sci. Univ. Tokyo, Section I* **7**, 463 (1957).
40. K. Yosida, *Lectures on Semi-Group Theory and Its Application to Cauchy's Problem in Partial Differential Equations*, Tata Institute of Fundamental Research, Bombay, India (1957).

Recent Mathematical Developments in Cascade Theory

S. K. SRINIVASAN*

INDIAN INSTITUTE OF TECHNOLOGY
Madras, India

1. INTRODUCTION

The cascade theory of cosmic ray showers originally suggested by Bhabha and Heitler[1] and Carlson and Oppenheimer has been developed to a considerable degree of sophistication.[2] As is often the case at this point, our understanding of the physical situation with respect to cascade theory is not satisfactory, but rather is so unsatisfactory that the problem demands immediate solution. To this end, I shall attempt to develop an electromagnetic cascade theory in the present paper.

As soon as the mean behavior of cascades was explained by the Bhabha Heitler theory, attempts were made to explain the fluctuation of the number distribution about the mean. At a time when this problem was considered almost insoluble, two different approaches were suggested. One approach consists of defining certain correlation functions (called "product densities" by Bhabha[3] and Ramakrishnan,[4] "cumulant functions" by Kendal,[5] and "density functions" by Janossy[6]) and relating them in turn to the moments of the number distribution. The other method, which is actually more useful, was suggested by Janossy[7]; it consists of defining the function $\pi_i(n, E, E_0, t)$ as the probability that n electrons are found at t cascading into energy states greater than E in a shower produced by a primary of the i-th type ($i = 1$ and $i = 2$ correspond to an electron- and photon-initiated shower, respectively). Using $\pi_i(n, E, E_0, t)$ and analyzing the various

---

* Department of Mathematics.

possible outcomes of events in the interval $(0, \Delta)$ of t, Janossy proved that π_i satisfies the equation

$$\frac{\partial \Pi_i(n, E, E_o, t)}{\partial t} = - \int_0^{E_0} R_i(E', E_0)dE' \Pi_i(n, E, E_0, t)$$

$$+ \sum_{n_1+n_2=n} \int_0^{E_0} R_i(E', E_0)dE' \Pi_i(n_1, E, E', t)\Pi_{3-i}(n_2, E, E_0 - E', t)$$

$$(1.1)$$

Equation (1.1) has been intensively studied with regard to the moments of the number distributions of electrons. However, because the equation is nonlinear, the problem was considered so difficult that few attempts have been made to use equation (1.1), or the moments derived from it, directly. Therefore, all work subsequent to that of Janossy is limited to mean behavior or, at most, the square deviation on the basis of the density functions.

2. A NEW APPROACH TO CASCADE THEORY

Based on the investigations of anomalous electron showers which followed the report of Schein et al., in 1954, Fay at Gottingen has pointed out (in a private communication) that it would be more convenient, especially for showers involving small thicknesses, to count electrons with specific reference to the energy level at the time they are produced. Thus, we can define $\pi_i(n, E, E_0, t)$ as the probability that n electrons are produced between 0 and t, the primitive energy of each of the electrons being greater than E. Encouraged by the usefulness of this approach, workers[8-10] have used the product density function as a basis for deriving results pertaining to the mean and mean square number. Srinivasan et al.[10] have shown that the mean square number of electrons produced between 0 and t can be expressed as a double Mellin integral with the integrand containing terms involving Mellin transforms of the product density of degree two photons and a mixed product density of degree two electrons and photons at t. Bhabha and Ramakrishnan[11] have obtained an explicit Mellin-transform solution of these product densities. However, since these expressions involve an enormous number of terms, the numerical evaluation of the Mellin integral is quite a difficult task. Actually, the mean square number of electrons and photons based on the results of Ramakrishnan and Srinivasan[8] has been calculated only for fairly large thicknesses,[11] where several terms differing by an order of magnitude can be neglected.

Evaluation of small thicknesses is still a problem in spite of the speed of modern computing devices. In this paper, we shall attempt to show that the difficulty can be overcome by dealing with π directly instead of associating its moments with the so-called density functions of different orders.

Let $\pi_i(n, E, E_0, t)$ be the probability that electrons are produced between 0 and t by a primary of energy E. As usual, we shall assume that the probability per unit thickness of matter that an electron of energy E will radiate a quantum and drop to an energy between E' and $E' + dE'$ is $R_1(E'|E)\,dE'$, and that the probability that a photon of energy E will be annihilated into an electron-positron pair, one of which has an energy between E' and $E' + dE'$, is $R_2(E'|E)\,dE'$. When screening is complete, R_1 and R_2 are given by[13]

$$R_1(E'|E) = \left[\frac{E - E'}{E} - (\tfrac{4}{3} + \alpha)\left(i - \frac{E}{E - E'}\right)\right]\frac{1}{E} \qquad (2.1)$$

$$R_2(E'|E) = \left[1 - (\tfrac{4}{3} + \alpha)\left(\frac{E'}{E} - \frac{E'^2}{E^2}\right)\right]\frac{1}{E} \qquad (2.2)$$

We shall take only these two processes into account and neglect collision loss. Because of the homogeneous nature of cross sections (2.1) and (2.2), $\pi_i(n, E, E_0, t)$ is a function only of E/E_0. Writing the function as $\pi_i(n, \epsilon; t)$ where $\epsilon = E/E_0$ we obtain

$$\frac{\partial \pi_0(n, \epsilon; t)}{\partial t} = -\pi_1(n, \epsilon; t)\int_0^t R_1(\epsilon')d\epsilon'$$

$$+ \sum_{m=0}^{\infty}\int_0^1 R_1(\epsilon')\pi_1\left(m, \frac{\epsilon}{\epsilon'}; t\right)\pi_2\left(n - m, \frac{\epsilon}{1 - \epsilon'}, t\right)d\epsilon' \quad (2.3)$$

$$\frac{\partial \pi_2(n, \epsilon; t)}{\partial t} = -\pi_2(n, \epsilon; t)\int_0^1 R_1(\epsilon')d\epsilon'$$

$$+ \int_0^\epsilon R_2(\epsilon')\pi_2\left(n - 1, \frac{\epsilon}{1 - \epsilon'}, t\right)d\epsilon'$$

$$+ \sum_{m=0}^{\infty}\int_\epsilon^{(1-\epsilon)} R_2(\epsilon')\pi_1\left(m, \frac{\epsilon}{\epsilon'}; t\right)\pi_2\left(n - m - 2, \frac{\epsilon}{1 - \epsilon'}, t\right)d\epsilon'$$

$$+ \int_{(1-\epsilon)}^1 R_2(\epsilon')\pi_1\left(n - 1, \frac{\epsilon}{\epsilon'}; t\right)d\epsilon' \qquad (2.4a)$$

with the initial conditions

$$\pi_1(n, \epsilon, 0) = \pi_2(n, \epsilon, 0) = \delta_0^n \qquad (2.5)$$

We observe that (2.3) holds true for the entire range $0 \leqslant \epsilon \leqslant 1$, while (2.4a) is valid only for the range $0 \leqslant \epsilon \leqslant \frac{1}{2}$. For $\epsilon > \frac{1}{2}$, π_2 satisfies the equation

$$\frac{\partial \pi_2(n, \epsilon, t)}{\partial t} = -\pi_2(n, \epsilon, t) \int_0^1 R_2(\epsilon') d\epsilon'$$

$$+ \delta_0^n \int_{(1-\epsilon)}^{\epsilon} R_2(\epsilon') d\epsilon'$$

$$+ \int_0^{(1-\epsilon)} R_2(\epsilon') \pi_1\left(n - 1, \frac{\epsilon}{1-\epsilon'}, t\right) d\epsilon'$$

$$+ \int_\epsilon^1 R_2(\epsilon') \pi_1\left(n - 1, \frac{\epsilon}{\epsilon'}, t\right) d\epsilon' \qquad (2.4b)$$

Comparing equations (2.3) and (2.4) with (1.1), we find that (2.3) is identical to the Janossy equation. The difference is brought out by (2.4a), where linear terms are integrated over partial ranges. This is due to the situation that at the regeneration point one or two electrons with energies above ϵ are produced. If $\epsilon' = E'/E_0$ falls within the interval $(\epsilon, 1 - \epsilon)$, then both the electrons produced have an energy above ϵ, and the second term on the right-hand side of (2.4a) corresponds to this situation. If however ϵ' falls outside the interval $(\epsilon, 1 - \epsilon)$, only one of the electrons has an energy above ϵ, and this is taken care of by the last two terms in (2.4a).

3. CASCADES FOR INFINITE THICKNESS

Denoting $\pi_i(n, E, E_0, t)$ as $\pi_i(n, \epsilon, t)$, where, as before, $\epsilon = E/E_0$, we set

$$\lim_{t \to \infty} \pi_i(n, \epsilon, t) = \pi_i(n, \epsilon) \qquad (3.1)$$

The existence of the limit is obvious from a physical point of view and has been discussed in reference 8. We note that in the limit as t tends to infinity, equations (2.3), (2.4a), and (2.4b) become

$$-\pi_1(n, \epsilon) \int_0^1 R_1(\epsilon') d\epsilon' + \sum_{m=0}^{\infty} \int_0^1 R_1(\epsilon') \pi_1\left(m, \frac{\epsilon}{\epsilon'}\right) \pi_1\left(n - m, \frac{\epsilon}{1-\epsilon'}\right) d\epsilon' = 0$$

$$(3.2)$$

$$\pi_2(n, \epsilon) \int_0^1 R_2(\epsilon') d\epsilon' = \int_0^{\epsilon} R_2(\epsilon') \pi_1\left(n - 1, \frac{\epsilon}{1 - \epsilon'}\right) d\epsilon'$$

$$+ \sum_{m=0}^{\infty} \int_\epsilon^{(1-\epsilon)} R_2(\epsilon') \pi_1\left(m, \frac{\epsilon}{\epsilon'}\right) \pi_1\left(n - m - 2, \frac{\epsilon}{1 - \epsilon'}\right) d\epsilon'$$

$$+ \int_{(1-\epsilon)}^1 R_2(\epsilon') \pi_1\left(n - 1, \frac{\epsilon}{\epsilon'}\right) d\epsilon' \qquad 0 \leqslant \epsilon \leqslant \frac{1}{2} \qquad (3.3a)$$

$$\pi_2(n, \epsilon) \int_0^1 R_2(\epsilon')d\epsilon' = \delta_0^n \int_{(1-\epsilon)}^\epsilon R_2(\epsilon')d\epsilon'$$

$$+ \int_0^{(1-\epsilon)} R_2(\epsilon')\pi_1\left(n - 1, \frac{\epsilon}{1 - \epsilon'}\right)d\epsilon'$$

$$+ \int_\epsilon^1 R_2(\epsilon')\pi_1\left(n - 1, \frac{\epsilon}{\epsilon'}\right)d\epsilon' \qquad \tfrac{1}{2} \leqslant \epsilon \leqslant 1 \qquad (3.3b)$$

Introducing the generating function $g_i(u, \epsilon)$,

$$g_i(u, \epsilon) = \sum_n \pi_i(n, \epsilon)u^n \qquad (3.4)$$

we obtain

$$g_i(u, \epsilon) \int_0^1 R_i(\epsilon')d\epsilon' = u^{i-1} \int_0^\epsilon g_{3-i}\left(u, \frac{\epsilon}{1 - \epsilon'}\right)R_i(\epsilon')d\epsilon'$$

$$+ u^{2i-2} \int_\epsilon^{(1-\epsilon)} g_i\left(u, \frac{\epsilon}{\epsilon'}\right)g_{3-i}\left(u, \frac{\epsilon}{1 - \epsilon'}\right)R_i(\epsilon')d\epsilon'$$

$$+ u^{i-1} \int_{(1-\epsilon)}^1 g_{3-i}\left(u, \frac{\epsilon}{\epsilon'}\right)R_i(\epsilon')d\epsilon' \qquad (3.5)$$

Equation (3.5) is valid for the entire range of ϵ for $i = 1$ only. When $i = 2$, equation (3.5) covers only the range $0 \leqslant \epsilon \leqslant \tfrac{1}{2}$. For $\epsilon > \tfrac{1}{2}$, we have

$$g_2(u, \epsilon) \int_0^1 R_2(\epsilon')d\epsilon' = \int_{(1-\epsilon)}^\epsilon R_2(\epsilon')d\epsilon'$$

$$+ u \int_0^{(1-\epsilon)} R_2(\epsilon')g_1\left(u, \frac{\epsilon}{1 - \epsilon'}\right)d\epsilon' + u \int_\epsilon^1 R_2(\epsilon')g_1\left(u, \frac{\epsilon}{\epsilon'}\right)d\epsilon' \qquad (3.6)$$

Equations (3.5) and (3.6) are not capable of explicit solution. However, if we restrict ourselves to the factorial moments, the problem becomes tractable. Defining the factorial moments as

$$N_m(\epsilon) = \frac{\partial^m g_1}{\partial u^m}\bigg|_{u=1}$$

$$M_m(\epsilon) = \frac{\partial^m g_2}{\partial u^m}\bigg|_{u=1} \qquad (3.7)$$

We find that the second factorial moments satisfy the equations

$$N_2(\epsilon) \int_0^1 R_1(\epsilon')d\epsilon' = \int_\epsilon^1 N_2\left(\frac{\epsilon}{\epsilon'}\right)R_1(\epsilon')d\epsilon' + \int_0^{(1-\epsilon)} M_2\left(\frac{\epsilon}{1 - \epsilon'}\right)R_1(\epsilon')d\epsilon'$$

$$+ 2[1 - H(\epsilon - \tfrac{1}{2})]\int_\epsilon^{(1-\epsilon)} N_1\left(\frac{\epsilon}{\epsilon'}\right)M_1\left(\frac{\epsilon}{1 - \epsilon'}\right)R_1(\epsilon')d\epsilon' \qquad (3.8)$$

$$M_2(\epsilon) \int_0^1 R_2(\epsilon')d\epsilon' = \int_0^{(1-\epsilon)} N_2\left(\frac{\epsilon}{1-\epsilon'}\right) R_2(\epsilon')d\epsilon'$$

$$+ \int_\epsilon^1 N_2\left(\frac{\epsilon}{\epsilon'}\right) R_2(\epsilon')d\epsilon' + 2\int_0^{(1-\epsilon)} N_1\left(\frac{\epsilon}{1-\epsilon'}\right) R_2(\epsilon')d\epsilon'$$

$$+ 2\int_\epsilon^1 N_1\left(\frac{\epsilon}{\epsilon'}\right) R_2(\epsilon')d\epsilon'$$

$$+ 2[1 - H(\epsilon - \tfrac{1}{2})]\int_\epsilon^{(1-\epsilon)} \left\{ N_1\left(\frac{\epsilon}{\epsilon'}\right) + M_1\left(\frac{\epsilon}{1-\epsilon'}\right) + 1 \right.$$

$$\left. + N_1\left(\frac{\epsilon}{\epsilon'}\right) M_1\left(\frac{\epsilon}{1-\epsilon'}\right) \right\} R_2(\epsilon')d\epsilon' \tag{3.9}$$

The first factorial moments are just the mean numbers and are given by (see, for example, reference 2)

$$N_1(\epsilon) = \frac{1}{2\pi i}\int_{(\sigma - i\infty)}^{(\sigma + i\infty)} \frac{B(s)C(s)}{(s-1)[A(s)D - B(s)C(s)]} e^{Y(S-1)} ds \tag{3.10}$$

$$M_1(\epsilon) = \frac{1}{2\pi i}\int_{(\sigma - i\infty)}^{(\sigma + i\infty)} \frac{B(s)A(s)}{(s-1)[A(s)D - B(s)C(s)]} e^{Y(S-1)} ds \tag{3.11}$$

where

$$y = \log \frac{E_0}{E} \tag{3.12}$$

$$A(s) = (\tfrac{4}{3} + \alpha)[\psi(s) + \gamma - 1] + \tfrac{1}{2} - \frac{1}{s[s+1]}$$

$$B(s) = 2\left[\frac{1}{s} - (\tfrac{4}{3} + \alpha)\frac{1}{(s+1)(s+2)}\right]$$

$$C(s) = \frac{1}{s+1} + (\tfrac{4}{3} + \alpha)\frac{1}{s(s-1)}$$

$$D = \tfrac{7}{9} - \tfrac{1}{8}\alpha \tag{3.13}$$

The α is a constant equal to 0.0246 and γ is the Euler–Macheroni constant.

To solve (2.8) and (2.9) we use the Mellin transform technique. Defining $N_2(s)$ and $M_2(s)$ by

$$N_2(s) = \int_0^1 N_2(\epsilon)\epsilon^{s-1} d\epsilon \tag{3.14}$$

$$M_2(s) = \int_0^1 M_2(\epsilon)\epsilon^{s-1} d\epsilon \tag{3.15}$$

we obtain

$$A(s+1)N_2(s) - C(s+1)M_2(s) = L_1(s+1) \tag{3.16}$$

$$-B(s+1)N_2(s) + DM_2(s) = L_2(s+1) \qquad (3.17)$$

where

$$L_1(s+1) = 2\int_0^{1/2} \epsilon^{s-1}\, d\epsilon \int_\epsilon^{(1-\epsilon)} N_1\left(\frac{\epsilon}{\epsilon'}\right) M_1\left(\frac{\epsilon}{1-\epsilon'}\right) R_1(\epsilon')\, d\epsilon' \qquad (3.18)$$

$$L_2(s+1) = 4\int_0^1 \epsilon^{s-1}\, d\epsilon \int_\epsilon^1 N_1\left(\frac{\epsilon}{\epsilon'}\right) R_2(\epsilon')\, d\epsilon'$$

$$+ 2\int_0^{1/2} \epsilon^{s-1}\, d\epsilon \int_\epsilon^{(1-\epsilon)} \left[N_1\left(\frac{\epsilon}{1-\epsilon'}\right) + M_1\left(\frac{\epsilon}{1-\epsilon'}\right) \right.$$

$$\left. + 1 + N_1\left(\frac{\epsilon}{\epsilon'}\right) M_1\left(\frac{\epsilon}{\epsilon'}\right) \right] R_2(\epsilon')\, d\epsilon' \qquad (3.19)$$

Thus, $N_2(\epsilon)$ and $M_2(\epsilon)$ can be explicitly solved once $N_1(\epsilon)$ and $M_1(\epsilon)$ are explicitly obtained.

4. SOLUTION OF THE MEAN NUMBERS

We now turn to the evaluation of the contour integrals given by (3.10) and (3.11). The integrand has no singularities to the right of the line $s = 2$ in the complex s-plane. At $s = 2$, there is a simple pole due to the zero of $\phi(s) = A(s)\, D - B(s)\, C(s)$. An examination of equations (3.13) shows that there are possible poles at $s = 1, 0, -1$, and -2 arising from the factor $B(s)\, C(s)$. In addition, there are poles arising from the possible zeros of $\phi(s)$.

Let us first evaluate the integral on the right-hand side of (3.10). Since $A(s)$ is a linear combination of $\psi(s)$, the digamma function* and $1/s(s+1)$, $A(s)$ has a series of poles at all negative integral values of s. Hence, it is easy to see that

$$\lim \frac{B(s)C(s)}{\phi(s)}$$

exists when s tends to any one of the points $0, -1$, and -2. Thus, the integrand has a simple pole at $s - 1$ due to the factor $(s - 1)$ in the denominator and another simple pole at $s = 2$ due to the simple zero of $\phi(s)$. By consideration of the signs of $B(s)\, C(s)$ and $A(s)$ it is easy to prove that $\phi(s)$ has no zeros on the portion of the real axis extending from -3 to 2. From -3 onward, there is a zero of $\phi(s)$ in the interval $(-n - 1, -n)$ (n being an integer), this being due to $\psi(s)$ taking all values from $-\infty$ to $+\infty$ in the interval $(-n - 1, -n)$. Apart from this,

---

* The $\psi(s)$ is defined to be $d/ds\,(\log s!)$.

there may be zeros of $\phi(s)$ off the real axis. However, it is shown in the next section that $\phi(s)$ has six zeros off the real axis.

Choosing $\sigma = \sigma_0 > 2$, we can evaluate the Mellin integral. Let us consider a typical pole at $s = +\lambda m$ on the negative real axis $(-m - 1 < \lambda_m < -m)$. The residue R_m from λ_m is given by

$$R_m = \lim_{s \to \lambda_m} (s - \lambda_m) \frac{B(s)C(s)e^{\Gamma(s-1)}}{\phi(s)(s - 1)} \tag{4.1}$$

Since

$$\lim \frac{s - \lambda_m}{\phi(s)} = \lim \frac{1}{[\phi(s) - \phi(\lambda_m)]/(s - \lambda_m)}$$

$$= \frac{1}{\phi'(\lambda_m)} \tag{4.2}$$

we find

$$R_m = \frac{e^{\Gamma(\lambda_m-1)}}{\lambda_m - 1} \frac{B(\lambda_m)C(\lambda_m)}{D A'(\lambda_m) - B'(\lambda_m)C(\lambda_m) - B(\lambda_m)C'(\lambda_m)} \tag{4.3}$$

We note that $|B(s)\ C(s)|$ is uniformly bounded in the strip $-m < s < -2$, and hence we obtain

$$|R_m| < e^{\Gamma(\lambda_m-1)} \frac{M}{|A'(\lambda_m) + \delta|} \tag{4.4}$$

where M and δ are some positive numbers less than unity. We next invoke the following useful property of the digamma function (see, for example, Jahnke and Emde[13]):

$$\left| \frac{10}{\psi'(s)} \right| \leqslant 1 \qquad \text{for } s \text{ real} \tag{4.5}$$

Using the above inequality, we note that

$$|R_m| < \frac{e^{\Gamma(\lambda_m-1)}}{10}$$

and hence

$$|\sum_{l=1}^{\infty} R_l| < \frac{e^{-\Gamma}}{10(1 - e^{-\Gamma})} \tag{4.6}$$

Thus, $N_1(\epsilon)$ can be very well approximated by the sum of the residues from the poles at $s = 1$ and 2, provided we can neglect the contribution from the four zeros of $\phi(s)$ lying off the real axis. We have not been able to establish mathematically that the contribution is negligible. However, the close agreement of the mean numbers arising from the poles on the real axis with those calculated numerically shows that the zeros of $\phi(s)$

may be in the half-plane real $s < 0$. Under this approximation $N_1(\epsilon)$ and $N_2(\epsilon)$ are given by

$$N_1(\epsilon) = \beta e^{1^r} - 1 = \beta\epsilon - 1$$
$$\beta = 0.443 \tag{4.7}$$

In the same way, $M_1(\epsilon)$ can be approximated, and we find

$$M_1(\epsilon) = \beta(\epsilon) \tag{4.8}$$

In Table I, we present the values of N_1 and M_1 as given by (4.7) and (4.8) as well as those computed earlier numerically.* The agreement with the numerically computed values shows the dominant nature of the poles at $s = 2$.

Table I

Y	N_1	N_1*
4	23.18	23.15
5	64.71	64.71
6	177.6	177.6
7	484.6	484.6
8	1319	131ɟ
9	3587	3587
10	9752	9752
11	2651×10	2651×10
12	7206×10	7206×10
13	1959×10^2	1959×10^2
14	5325×10^2	5325×10^2
15	1447×10^3	1447×10^3

* Obtained by numerical evaluation of the inversion integral.

5. ZEROS OF THE FUNCTION $\phi(s)$

We shall examine the zeros of $\phi(s)$ lying off the real axis in the complex s-plane. We recall that $\psi(s)$ has the asymptotic expression for $|s| \gg 1$ arg $s \neq \pm\pi$ (see for example, reference 13):

$$\psi(s) = \log s + \frac{1}{2s} - \frac{1}{12s^2} - \frac{1}{120s^4} \tag{5.1}$$

The $\psi(s)$ is well represented by (5.1) for $|s| \gg 1$. Thus, we rewrite $\phi(s)$ as

* These values were evaluated in 1957 by one of us (S.K.S.) with the help of SILLIAC, the electronic digital computer at the School of Physics, University of Sydney.

$$\phi(s) = (\tfrac{4}{3} + \alpha)D\psi(s) - 0.0497 - \frac{D}{s(s+1)} - B(s)C(s)$$
$$= g(s) + h(s) \tag{5.2}$$

where

$$g(s) = (\tfrac{4}{3} + \alpha)D\psi(s) \tag{5.3}$$

With the help of (5.1), it is easy to see that

$$|g(s)| > |h(s)| \tag{5.4}$$

for $|s| \gg 1$ and arg $s \neq \pm\pi$.

Let us consider the contour shown in Fig. 3. A is the point whose s-value is $-N + \delta$ where δ is a small positive number. We shall choose N arbitrarily large. E is point $(3, 0)$ in the Argand s-plane. Under these conditions, it is easy to verify that

$$|g(s)| > |h(s)|$$

on the contour $A\ B\ C\ D\ E\ D'\ C'\ B'\ A$ (see Fig. 1). Since $g(s)$ and $h(s)$ are meromorphic functions of $s = p$, it follows by Roche's theorem (see, for example, Hille[14]) that the difference between the number of zeros and poles of $g(s)$ included in the region enclosed by the contour of Fig. 3 is the same as that of $\phi(s)$. From the properties of the digam-

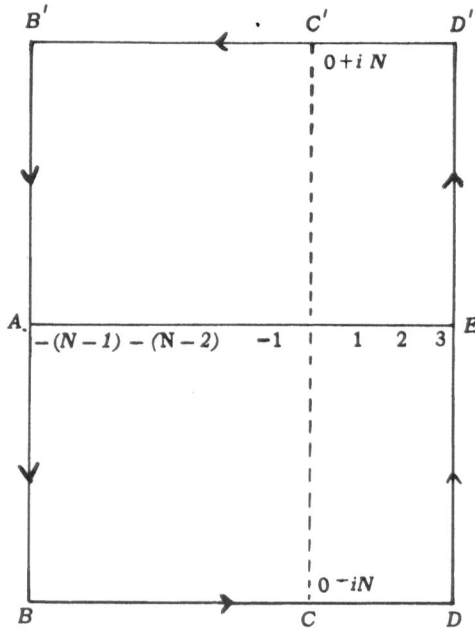

Fig. 1

ma function, it is clear that the difference is one. However, on the portion of the real axis included in the contour there are poles at negative integral values of s as well as at $s = 0$ and 1. All these poles are simple except at $s = 0$ and -1, which are second order poles for $\phi(s)$. In addition, $\phi(s)$ has simple zeros in each of the intervals $(-m, -m + 1)$, m being > 3 and there being no further zeros in the interval $(-3, 2)$ except at the point 2, where $\phi(s)$ has a simple zero. Thus, the difference between the number of zeros and poles of $\phi(s)$ is -5. Since the difference is $+1$ for $g(s)$, $\phi(s)$ has *six* zeros off the real axis. From the asymptotic properties of $\psi(s)$, it can be shown that the zero should lie in a region limited by the circle $|z| = r$, r, being of the order of unity. However, the agreement of the mean numbers obtained by neglecting the contribution from these zeros with those obtained by the saddle-point method as well as by numerical contour integration shows that the zeros should lie sufficiently left of the line Re $s = 1$.

Locating these six zeros of $\phi(s)$ appears to be difficult and is being attempted by numerical methods. By the Schwarz reflection principle, there should be three zeros with a positive imaginary part and three with a negative imaginary part. One zero of $\phi(s)$ is found to lie between $0.1 + 10.9$ and $1 + i\,0.8$.

6. HIGHER MOMENTS OF THE NUMBER DISTRIBUTION

Once we have explicit expressions, it is not difficult to obtain explicit expressions for the higher moments of the distribution. We shall summarize in this section the main results that can be derived. Denoting by $N_i(\epsilon)$ and $M_i(\epsilon)$ the i-th factorial moments of the number of electrons produced by the electron and photon primary, respectively, we obtain

$$N_2(\epsilon) = 0.2102\,e^{2Y} + 0.2134\,e^{Y} + 2.2534 \tag{6.1}$$

$$M_2(\epsilon) = 0.2420\,e^{2Y} + 0.2134\,e^{Y} \tag{6.2}$$

$$N_3(\epsilon) = 0.2802\,e^{3Y} + 0.2178\,e^{2Y} - 0.6430\,e^{Y} \tag{6.3}$$

$$M_3(\epsilon) = 0.1921\,e^{3Y} + 0.8659\,e^{2Y} - 0.6430\,e^{Y} \tag{6.4}$$

In cascade theory it is customary to compare the mean square deviation with the deviation corresponding to Poisson and Furry distributions (see, for example, Ramakrishnan[15]). Figures 2 and 3 demonstrate the variation of σ_1^2/σ_{1p}^2, σ_2^2/σ_{2p}^2, σ_1^2/σ_{1F}^2, and σ_2^2/σ_{2F}^2 with Y.

The possibility of explicit solutions of the moments for the special case of infinite thickness gives us hope that it may be possible to

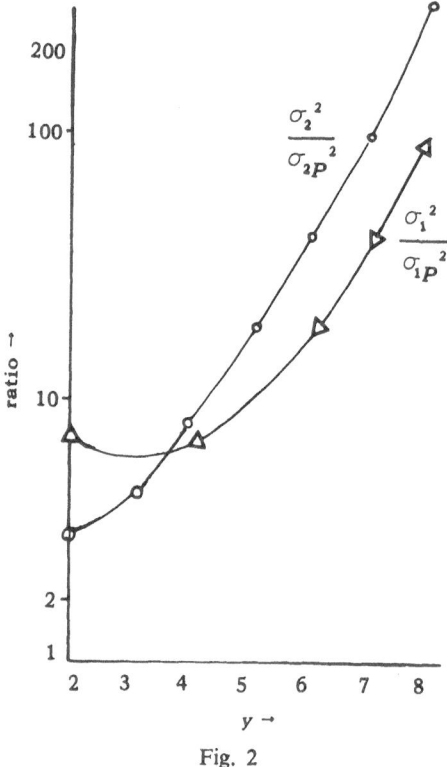

Fig. 2

obtain an explicit solution for cascades of finite thicknesses. In any case, it is our hope that for small thicknesses, expansion of the exponential functions of $t\lambda(s)$ and $t\mu(s)$ [see equation (19) of reference 1] may yield some fruitful results. The occurrence of branch cuts for $\lambda(s)$ and $\mu(s)$ should not be a serious handicap, and it is our conjecture that a rearrangement of the infinite series may completely eliminate the terms having branch cuts. If this be the case, the fluctuations are determined by a very simple law which can be used in conjunction with experimental data. Besides, a knowledge of the fluctuations would enable us to predict the role of the multiple processes such as tridents and multiple production of photons in highly energetic electromagnetic cascades.

NOTE

After the lecture notes were finalized, we obtained explicit analytical solutions for the mean square deviation for small thicknesses. Besides,

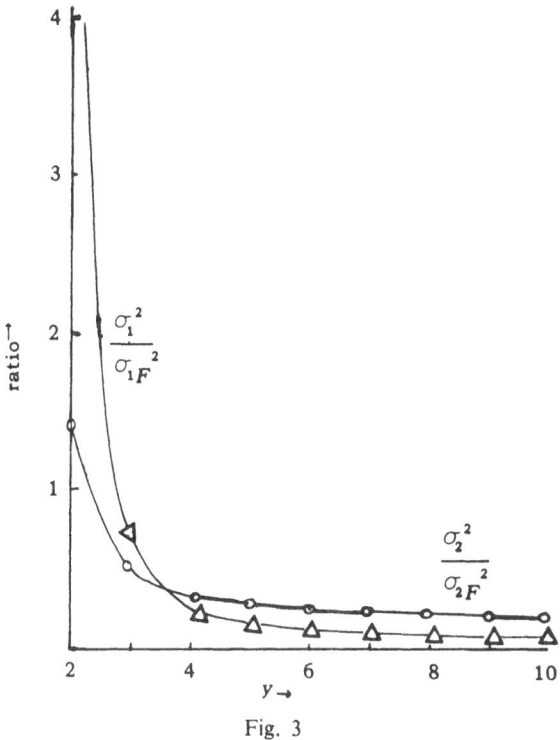

Fig. 3

the solutions have been extended to the case when the ionization loss is also material. For details the reader is referred to a forthcoming paper by S.K. Srinivasan and K.S.S. Iyer in *Ztschr. Phys.* (1964) (in press).

REFERENCES

1. H.J. Bhabha and W. Heitler, *Proc. Roy. Soc.* **159**, 432 (1937). J.F. Carlson and J.R. Oppenheimer, *Phys. Rev.* **51**, 220 (1937).
2. For an extensive review of the methods, see Ramakrishnan and Mathews, *Progr. Theoret. Phys.* **11**, 95 (1954).
3. H.J. Bhabha, *Proc. Roy. Soc. London, Ser. A* **202**, 301 (1950).
4. A. Ramakrishnan, *Proc. Roy. Soc. London, Ser. A* **202**, 301 (1950).
5. D.G. Kendal, *Proc. Roy. Stat. Soc.* **11 B**, 230 (1949).
6. L. Janossy, *Proc. Roy. Irish Acad. Sci.* **53 B**, 181 (1950).
7. L. Janossy, *Proc. Phys. Soc.* **63 A**, 241 (1950).
8. A. Ramakrishnan and S.K. Srinivasan, *Proc. Indian Acad. Sci.* **A 44**, 44 (1956).
9. S.K. Srinivasan and N.R. Ranganathan, *Proc. Indian Acad. Aci.* **A 45**, 69, 268 (1957).

10. S.K. Srinivasan, J.C. Butcher, B.A. Charthes, and H. Mesel, *Nuovo Cimento* **9**, 77 (1958).
11. A. Ramakrishnan and S.K. Srinivasan, *Progr. Theoret. Phys.* **13**, 95 (1955).
12. H.A. Bethe and W. Heitler, *Proc. Roy. Soc. London, Ser. A* **146**, 83 (1934).
13. E. Jahnke and F. Emde, *Tables of Functions*, Dover, New York (1945), p. 19.
14. E. Hille, *Analytic Function Theory, Vol. 1*, Ginn and Co., New York (1959), p. 254.
15. A. Ramakrishnan, *Theory of Elementary Particles and Cosmic Rays*, Pergamon Press, New York (1962), p. 450.

Theory of a General Quantum System Interacting with a Linear Dissipation System

R. VASUDEVAN

MATSCIENCE
Madras, India

In this paper, we will try to sketch a formalism developed in a paper by Feynman and Vernon,[1] using Feynman's space–time formulation of nonrelativistic quantum mechanics. It might be quite useful if we could calculate, in terms of its own variables only, the behavior of a system of interest which is coupled to other external quantum systems. The purpose of this paper is to show that in certain cases the effect of the external systems can be included in a class of functionals (called *influence functionals*) of the coordinates of the system itself.

The origins of the present paper[2] lie in the classic paper of Feynman in 1948, where the Schrödinger equation was derived using the path-integral formalism. Of course, this formulation contains the concept of probability amplitude, associated with a completely specified motion as a function of time. To understand this we have to bear in mind the following basic ideas (and note their important distinction from the more common ideas of the usual probability) which lead to the space-time formulation of quantum mechanics; the most important of them are given in detail below.

1. SUPERPOSITION OF PROBABILITIES IN QUANTUM MECHANICS

If three events A, B, C occur and if P_{ab} is the probability that if A gives the result a then B gives the result b, and if P_{bc} is the probability that if B gives the result b then C gives c, and if the two are independent events, then

$$P_{ac} = \sum_b P_{ab} P_{bc} \tag{1}$$

In quantum mechanics there exist complex numbers φ_{ab} such that $P_{ab} = |\varphi_{ab}|^2$ and $P_{bc} = |\varphi_{bc}|^2$. However, if *no attempt is made* to measure B in quantum mechanics, the equation (1) of classical probability theory is to be replaced by

$$\varphi_{ac} = \sum_b \varphi_{ab} \varphi_{bc} \tag{2}$$

2. PROBABILITY AMPLITUDES OF A GIVEN PATH

A. Suppose a particle occupies different points $x_1 \ldots x_n$ at times $t_1 \ldots t_n$; the probability of such a path is, say, $P(x_1, x_2, \ldots, x_n)$. The probability that such a path lies in a given region is obtained by integrating $x_1, x_2, \ldots, x_n$ over that region for allowed values of x_1 $x_2, \ldots, x_n$:

$$\int\int_{a_2}^{b_2} \int_{a_1}^{b_1} P(x_1, x_2, \ldots, x_n)\, dx_1\, dx_2 \cdots dx_n$$

If the particle takes any one such path in a given region R, according to quantum-mechanical concepts the process can be described by a probability amplitude φ_R:

$$\varphi_R = \lim_{\epsilon \to 0} \int_R \Phi(x_1, \ldots, x_i x_{i+1}) \cdots dx_i dx_{i+1} \tag{3}$$

where $\Phi(x_1 \ldots x_i(t) \cdots)$ is the complex contribution from each path and a path is a sequence of configurations for successive times t_i, $t_{i+1} \ldots$, etc., where $t_{i+1} = t_i + \epsilon$ and the limit $\epsilon \to 0$ should be taken at the end of the calculation.

B. The second postulate tells how to compute the important quantity Φ for each path.

The paths contribute equally in magnitude, but the phase of their contribution is the classical action (in units of $\hbar$), that is, the time integral of the Lagrangian taken along the path, which means that $\Phi[x(t)]$ is a functional of the path $x(t)$ and is proportional to

$$\Phi[x(t)] \propto \exp\frac{i}{\hbar} S[x(t)] \tag{4}$$

where $S[x(t)] = \int L[\dot{x}(t), x(t)]\, dt$ is the time integral of the Lagrangian taken along the path considered. And in going from x_i to x_{i+1}, we take such an L, where the action is minimized:

$$S(x_i, x_{i+1}) = \text{Min} \int_{t_i}^{t_{i+1}} L[\dot{x}(t), x(t)]\, dt$$

and

$$S = \sum_i S_i(x_i, x_{i+1}) \tag{5}$$

This is equivalent to saying that we take the Lagrangian corresponding to the classical path and that L is a function only of x and $\dot{x}(t)$:

$$\varphi(R) = \lim_{\epsilon \to 0} \int_R \exp \frac{i}{\hbar} \left[\sum_i S(x_{i+1}, x_i)\right] \cdots \frac{dx_{i+1}}{A} \frac{dx_i}{A} \tag{6}$$

where we have let the normalization factor be split into a factor $1/A$ for each instant. The integration is over all such points x_i which lie in the region R. The physical interpretation that $|\varphi(R)|^2$ is the probability that the particle will be found in R is a vital statement in the present formulation of quantum mechanics.

3. THE WAVE FUNCTION

If a particle is in a certain region of space–time, say, R', it may said to be defined by a wave function $\psi(x, t)$ determined only by its past and containing all that is needed to determine its future. The above formulation then leads to the result that the chance of a system in state ψ being found in state χ is, according to the first principles evolved here, given by

$$\left| \int \chi^*(x, t)\, \psi(x, t)\, dx \right|^2 \tag{7}$$

The wave function at x'' and t'' can be obtained from the wave function at x' and t' by

$$\psi(x'', t'') =$$

$$\lim_{\epsilon \to 0} \int \cdots \int \exp\left[\frac{i}{\hbar} \sum_{i=0}^{J-1} S(x_{i+1}, x_i)\right] \psi(x', t') \frac{dx_0}{A} \frac{dx_1}{A} \cdots \frac{dx_{j+1}}{A} \tag{8}$$

where $x_0 = x'$, $x_j = x''$, and $j\epsilon = t'' - t'$. It is easy to see that only the paths quite close to the classical path can contribute strongly to the integral, since $\hbar$ is very near zero and the phase oscillates very rapidly, even for neighboring paths.

The Wave Equation

Let us relate $\psi(x_{k+1}, t + \epsilon)$ to $\psi(x_k, t)$ for a particle of mass m moving in one dimension under a potential $V(x)$. Then

$$S(x_{i+1}, x_i) = \epsilon L\left[\frac{x_{i+1} - x_i}{\epsilon}, x_{i+1}\right]$$

$$= \frac{m\epsilon}{2}\left[\frac{x_{i+1} - x_i}{\epsilon}\right]^2 - \epsilon V(x_{i+1})$$

Hence,

$$\psi(x_{k+1}, t + \epsilon)$$

$$= \int \exp \frac{i\epsilon}{\hbar}\left[\frac{m}{2}\left(\frac{x_{k+1} - x_k}{\epsilon}\right)^2 - V(x_{k+1})\right]\psi(x_k\, t_k)\frac{dx_k}{A}$$

Let

$$x_{k+1} = x \qquad x_{k+1} - x_k = \xi \qquad x_k = (x - \xi)$$

$$\psi(x, t + \epsilon) = \int \exp \frac{im\xi^2}{\epsilon 2\hbar} \exp \frac{-i\epsilon\, V(x)}{\hbar}\, \psi(x - \xi, t)\frac{d\xi}{A} \tag{9}$$

We can expand $\psi(x - \xi, t)$ in a Taylor series, since the integral contributes only for small ξ. We keep terms up to order ξ^2 in the expansion of ψ. Also, we expand ψ on the left-hand side to first order in ϵ. To make both sides agree in zero order in ϵ, we have to put $A = (2\pi \hbar \epsilon i/m)^{1/2}$, and we obtain the Schrödinger equation

$$\frac{\hbar}{i}\frac{\partial\psi}{\partial t} = \frac{1}{2m}\left(\frac{\hbar}{i}\frac{\partial}{\partial x}\right)^2 \psi + V(x)\psi \tag{10}$$

Matrix Elements

The chance that a system found in state ψ at x' and t' will afterward be found in a state χ at x'' and t'' is given by the square of the transition amplitude:

$$\langle \chi_{t''} | 1 | \psi_{t'} \rangle_s =$$

$$\lim_{\epsilon \to 0} \int \cdots \int \chi^*(x''\, t'')\left(\exp\frac{i}{\hbar}\, S\right)\psi(x't')\frac{dx_0}{A} \cdots \frac{dx_{j-1}}{A}\, dx_j \tag{11}$$

where $x_0 = x'$ and $x_j = x''$. If F is any function of the coordinates $x_0, \ldots, x_i, \ldots$ for $t' < t < t''$, we define the matrix element of F between state ψ at t' and state χ at t'' for action S as

$$\langle \chi_{t''} | F | \psi_{t'} \rangle_s =$$

$$\lim_{\epsilon \to 0} \int \cdots \int \chi_t^*(x'', t'')\, F(x_0 \cdots x_j)\left[\exp\frac{i}{\hbar}\sum_{i=0}^{j=1} S(x_{i+1}\, x_i)\right] \tag{12}$$

$$\times \psi(x', t')\frac{dx_0}{A} \cdots \frac{dx_{j-1}}{A}\, dx_j$$

And in the limit $\epsilon \to 0$, F is a function of the path $x(t)$.

To illustrate this, let us think of the interaction of two particles. This interaction can be represented by a field consisting of a set of oscillators. The equation of motion of the oscillators can be solved, and the oscillators eliminated. As a simple case, suppose a particle of coordinates x and Lagrangian $L(\dot{x}x)$ interacts with an oscillator with coordinates $q(t)$ and Lagrangian $\frac{1}{2}(\dot{q}^2 - \omega q^2)$ through a term $V(x, t)$ $q(t)$. Let the initial state of the particle be $\psi(t')$ at t' and φ_n be the oscillator with n-th energy-level wave function. Let the final state be $\chi_{t''}$ for the particle at time t'' and φ_m for the oscillator level m. The transition probability is given by

$$\langle \chi_{t''}\varphi_m \mid 1 \mid \psi_{t'}\varphi_n\rangle_{S_p+S_0+S_I} =$$

$$\int \cdots \int \varphi_m^*(q_{t''})\, \chi_{t''}(x_{t''}) \exp \frac{i}{\hbar}(S_p + S_0 + S_I) \tag{13}$$

$$\times\; \psi_t(x_0)\varphi_n(q_0)\frac{dx_0}{A}\frac{dq_0}{a}\cdots\frac{dx_{j-1}}{A}\frac{dq_{j-1}}{a}\,dx_j\,dq_j$$

S_p is the action

$$\sum_0^{J-1} S_p(x_{i+1}, x_i)$$

for the particle, and

$$S_0 = \sum_{i=0}^{J-1}\left[\frac{\epsilon}{2}\left(\frac{q_{i+1}-q_i}{\epsilon}\right)^2 - \frac{\epsilon\omega^2}{2}q_{i+1}^2\right] \tag{14}$$

for the oscillator, and

$$S_I = \sum_{i=0}^{J-1}\gamma_i(x_i)q_i$$

is the action corresponding to the interaction between the particle and the oscillator. The normalizing constant a for the oscillator is $(2\pi i\epsilon/h)^{1/2}$. Here, all the integrations can be easily performed over the q_i, $0 < i < j$. The result of these integrations is that we are left with

$$\left(2\pi i h\frac{\sin \omega T}{\omega}\right)^{-1/2}\exp\frac{i}{\hbar}\,[S_p + Q(q_j,q_0)] \tag{15}$$

where $T = t'' - t'$, and

$$Q(q_j,q_0) = \frac{\omega}{2\sin\omega T}\Bigg[(\cos\omega T)(q_j^2 + q_0^2) - 2q_jq_0$$

$$+ \frac{2q_0}{\omega}\int_{t'}^{t''}\gamma(t)\sin\omega(t-t')\,dt$$

$$+ \frac{2q_j}{\omega}\int_{t'}^{t''}\gamma(t)\sin\omega(t''-t')\,dt$$

$$- \frac{2}{\omega^2}\int_{t'}^{t''}\int_{t'}^{t''}\gamma(t)\gamma(s)\sin\omega(t''-t)\sin\omega(t-t')\,ds\,dt\Bigg]$$

Therefore,

$$\langle \chi_{t''} \varphi_m | 1 | \psi_{t'} \varphi_n \rangle_{S_p + S_0 + S_I}$$

$$= \int \cdots \int \chi_{t''}^*(x_j) G_{mn} \exp \frac{iS_p}{\hbar} \psi_{t'}(x_0) \frac{dx_0}{A} \cdots \frac{dx_{j-1}}{A} dx_j \qquad (16)$$

$$= \langle \chi_{t'} | G_{mn} | \psi \rangle_{S_p}$$

which now contains the coordinates of the particle only:

$$G_{mn} = \left(\frac{2\pi i\hbar \sin \omega T}{\omega} \right)^{1/2} \iint \varphi_m^*(q_j) \exp \left[\frac{iQ(q_j, q_0)}{\hbar} \right]$$

$$\times \varphi_n(q_0) \, dq_j dq_0 \qquad (17)$$

Thus, all the oscillator coordinates are eliminated and the particle coordinates have been suitably modified.

In deriving expression (15), we can take the oscillator Lagrangian as $L = \frac{1}{2}(\dot{q}^2 - \omega^2 q^2)$ and divide q into two parts $q = q_0(t) + \chi(t)$ where $q_0(t)$ is the solution of the classical equation $\ddot{q}_0 + \omega^2 q_0 = 0$. To satisfy the initial conditions such that $q(t') = q'$ and $q(t'') = q''$, we have for q:

$$q = \frac{1}{\sin \omega T}[q' \sin \omega(t - t'') + q'' \sin \omega(t' - t)]$$

Hence

$$\int L_c \, dt = \frac{1}{2} q_c \dot{q}_c \Big|_{t''}^{t'} = \frac{\omega}{2 \sin \omega T}[(q''^2 + q'^2) \cos \omega T - 2q'q''] \quad (18)$$

Now the total Lagrangian is $L = L_c + \frac{1}{2}(\dot{\chi}^2 - \omega^2 \chi^2)$ and the variation of the path really means variation of χ, the deviation from the classical path, which therefore leads to $\chi(t'') = \chi(t') = 0$. Here, χ can be expanded as $\chi = \sum a_n [\sin n\pi(t - t')]/T$, and the path integration really means integrating over da_n from $-\infty$ to $+\infty$. Doing this, one finds that for a free as well as a harmonic oscillator, the $K_a(q', q'', t', t'')$ is the same as that for the classical case. The result (15) is for a harmonic oscillator with an interaction of the form

$$L_I = -\gamma Q(X)q \qquad (19)$$

Our present interest is to investigate the quantum behavior of a system when it is coupled to one or more systems—say, measuring instruments.

Let us suppose there are two nonrelativistic quantum systems whose coordinates are represented in a general way by Q and X, coupled by an interaction potential which is a function of the Q and

X. The complete solution of the problem is the solution of the Schrödinger equation for the system:

$$[H(Q) + H(X) + V(Q, X)] \psi(Q, X) = -\frac{\hbar}{i} \frac{\partial}{\partial t} \psi(Q, X) \qquad (20)$$

and this is an extremely difficult problem. However, by use of the
Lagrangian formalism of Feynman, the effect of the interaction systems
on the test system can be evaluated in a very neat form. Certain functionals called the *influence functionals* are defined which contain the
effects of the interacting system (x) on the test system in terms of the
coordinates of the test system only. In the paper cited above, the program followed was: (1) General properties of these functionals and their
relation to statistical mechanics were considered. (2) The influence
functionals for the following interacting systems were considered:
(a) definite classical forces, (b) random classical forces, (c) linear
interacting systems at zero and higher temperatures, (d) linear systems
driven by classical forces, (e) weakly coupled systems and maser
noise. Spin was not considered in this program, but it is not an additional problem.

4. INFLUENCE FUNCTIONALS

Let a quantum system be denoted by a general coordinate Q_t at
time t. The amplitude for the system to go from Q_t at t to Q_T at T is
given, as explained earlier, by

$$K(Q_T, T; Q_t, t) = \int \exp\left[\frac{i}{\hbar} S(Q)\right] DQ(t) \qquad (21)$$

where

$$S = \int_{t'}^{T} L(\dot{Q}, Q, t) \, dt$$

L being the classical Lagrangian for the path $Q(t)$, which is the same
expression as given in equation (9). Only the paths very near the
classical path contribute. The probability amplitude for the system to
go from the state ϕ_n at Q_τ to the state Q_m at $t = T$ is given by

$$A_{mn} = \int \phi_m^*(Q_T) K(Q_T, Q_\tau) \phi_n(Q_\tau) \, dQ_T dQ_\tau$$

and the transition probability P_{mn} to go from $n \to m$ is given by
$|A_{mn}|^2$:

$$P_{mn} = \int \phi_m^*(Q_T)\phi_m(Q_T') \exp\left\{\frac{i}{\hbar}\left[S(Q) - S(Q')\right]\right\}$$

$$\times \phi_n(Q_\tau)\phi_n(Q_\tau')\,DQ(t)\,DQ'(t)\,dQ_\tau\,dQ_\tau'\,dQ_T\,dQ_T' \qquad (22)$$

Let us now take two systems Q and X coupled by a potential $V(Q, X)$. Let the initial states of the systems Q and X be $\phi_n(Q_\tau)$ and $\chi_P(X_\tau)$ and the final states be $\phi_k(Q_T)$ and $\chi_P(X_T)$. In this case,

$$A_{mfni} = \int \phi_m^*(Q_T)\chi_f^*(X_T) \exp\left[\frac{i}{\hbar}\,S(Q, X)\right]\phi_m(Q_T)$$

$$\times \chi_i(X_T)\,DQ(t)\,DX(t)\,dQ_\tau\,dX_\tau\,dQ_T\,dX_T \qquad (23)$$

$S(Q, X)$ represents the classical action of the entire system of both Q and X. If the integration over the variables X is done first, what is left is the A_{mn} for Q with only Q variables wherein the effects of X have been included.

5. DEFINITION OF $F(Q, Q')$

For a system Q acted on by external systems, the transition probability that it makes a transition from a state $\psi_n(Q_T)$ to $\psi_m(Q_T)$ is

$$P_{mn} = \int \psi_m^*(Q_T)\psi_m(Q_T') \exp\left\{\frac{i}{\hbar}\left[S_0(Q) - S_0(Q')\right]\right\} F(Q, Q')$$

$$\times \psi_n^*(Q_\tau')\psi_n(Q_\tau)\,DQ(t)\,DQ'(t)\,dQ_\tau\,dQ_\tau'\,dQ_T\,dQ_T' \qquad (24)$$

where $F(Q, Q')$ contains all the effects of external influences on Q and

$$S_0(Q) = \int_\tau^T L(\dot{Q}, Qt)\,dt$$

which is the action of Q without external disturbances. Let the external system X be initially in $\chi_i(X_t)$ and later at T in $\chi_f(X_T)$ and let $V(Q, X, t)$ be the coupling potential and $S(X)$ be the action for the external system:

$$P_{mf.ni} = |A_{mf\,ni}|^2$$

$$= \int \psi_m^*(Q_T)\psi_m(Q_T')\chi_f^*(X_T)\chi_f(X_T')$$

$$\times \exp i/\hbar\,[S_0(Q) - S_0(Q') + S(X) - S(X') + S_i(QX) - S_i(Q'X')]$$

$$\times \psi_m^*(Q_\tau')\psi_n(Q_\tau)\chi_i^*(X_\tau')\chi_i(X_\tau)\,d\chi_\tau\ldots dQ_\tau$$

$$\times DX(t')\,DX(t)\,DQ(t')\,DQ(t) \qquad (25)$$

If now the integrations over all coordinates other than Q and Q' are performed, we obtain

$$F(Q, Q') = \int \chi_f^*(X_T)\chi_f(X_T')$$

$$\times \exp\left\{\frac{i}{\hbar} S[(X) - S(X') + S_I(QX) - S_I(Q'X')]\right\} \qquad (26)$$

$$\times \chi_i^*(X_\tau)\chi_i(X_\tau')DX(t)DX'(t)\,dX_\tau\,dX_\tau'\,dX_T\,dX_T'$$

that is, in terms of kernels we have

$$F(Q, Q') = \int \chi_f^*(X_T)\chi_f(X_T')K_Q(X_T X_\tau)K_{Q'}(X_T' X_\tau')$$

$$\times \chi_i^*(X_\tau')\chi_i(X_\tau)dX_\tau \dots dX_T' \qquad (27)$$

F is a functional whose form depends on the system X, its initial and final states, and the interaction $V(Q, X)$; if F_A on $Q = F_B$ on Q, then A and B have the same effect on Q. The same form of F is appropriate when the interaction system is composed of a linear system or a combination of linear systems.

6. GENERAL PROPERTIES OF $F(Q, Q')$

A. If in a physical situation the initial or final state of the interacting system is not known precisely, but the probability weight for the p-th situation is ω_p and $F_p(Q, Q')$ is the influence functional for that situation, the effective F is

$$F = \sum_p \omega_p F_p = \langle F \rangle$$

and this is to be used in calculating

$$\sum_i \omega_i P_{mfni}$$

for transitions in Q.

B. If a number of statistically and dynamically independent partial systems act on Q at the same time and if F^k is the influence of the k-th system alone, then the total influence of all systems is given by the product of the individual $F^{(k)}$:

$$F = \prod_{k=1}^N F^k$$

C. It is convenient to write F in terms of Φ, called the *influence phase*, defined by $F = \exp i\Phi$, and for independent disturbing systems the phases simply add.

D. It is evident from equation (27) that

$$F^*(Q, Q') = F(Q', Q)$$

E. If the final state f of the interaction system X is arbitrary, we have to sum over all f:

$$\sum_f {}_fF_i(Q, Q') = F_i(Q, Q')$$

And also, if $Q(t) = Q'(t)$ for all t, then $F_i(Q, Q)$ becomes independent of Q. This can be seen if we write

$$F_i(Q,Q) = \int \sum_f \chi_f^*(X_T)\chi_f(X_T')$$

$$\times \exp \left\{ \frac{i}{\hbar} [S(X) - S(X') + S_I(QX) - S_I(QX)] \right\} \qquad (28)$$

$$\times \chi_i^*(X_\tau')\chi_i(X_T)dX_T \ldots DX'(t)$$

Since in the interaction term S_I, Q is the same when the system is X or X', Q loses its identity as the coordinate of a quantum system and may be just a number; that is, $S_I(Q, X)$ may be interpreted as a driving potential which takes the system in an initial state i to any final state. $F_i(Q, Q)$ is the total probability of jumping away from i, which is of course unity. Thus, $F_i(Q, Q) = 1$ and is independent of Q. If, however, $Q = Q'$ only for $t > r$, then $F_i(Q, Q')$ is independent of Q only for $t \gg r$.

7. STATISTICAL MECHANICS

The significance of the influence functionals is felt in calculating the density matrix $\rho(Q_T, Q_T')$ at time T from $\rho(Q_\tau, Q_\tau')$ at time τ after eliminating the coordinates of the interaction systems. Taking the coordinate representation of the density matrix for the test system and the interaction system as

$$\rho(Q, X, Q', X') = \langle \psi(QX)\psi^*(Q'X')_{Av} \rangle$$

where the averaging is over the ensemble, the expectation value of an operator A acting on Q variable is

$$\langle A \rangle = \int\int\int \rho(Q, X, Q', X')A(Q, Q')dQ\,dQ'\,dX$$

where

$$A(Q',Q) = \sum_{i,j} A_{ij}\phi_i^*(Q)\phi_j(Q')$$

and

$$A_{ij} = \int \phi_i^*(Q) A \phi_j(Q) dQ \tag{29}$$

$\phi_i(Q)$ being one of a set of complete orthonormal eigenfunctions. We want to derive

$$\int \rho(Q_T, X_T, Q_T', X_T') dX_T = \rho(Q_T, Q_T')$$

$$\rho(Q_T, X_T; \ Q_{T'}, X_{T'}) = \tag{30}$$

$$\int \exp\left\{\frac{i}{h}[S_0(Q) - S_0(Q') + S(X) - S(X') + S_I(QX) - S_I(Q'X')]\right\}$$

$$\times \rho(Q_\tau, X_\tau; \ Q_\tau', X_\tau') DQ(t) DQ'(t) DX(t) DX'(t) dQ_\tau dX_\tau dQ_\tau' dX_\tau'$$

This is obtained by just taking the propagation of the wave functions from Q_τ to Q_T and X_τ to X_T. If the two systems are initially independent,

$$\rho(Q_\tau, X_\tau; \ Q_\tau', X_\tau') = \rho(Q_\tau Q_\tau')\rho(X_\tau X_\tau') \tag{31}$$

then eliminating X_T by putting X_T to be X_T' and integrating over Q_τ, we have

$$\rho(Q_T, Q_T') = \int \left\{ S(X_T - X_T') \exp\left[\frac{i}{h}[S(X) - S(X') + S_I(QX) - S_I(Q'X')]\right] \right.$$

$$\times \rho(X_\tau X_\tau') DX(t) DX'(t) dX_\tau dX_\tau' dX_\tau' dX_T \right\} \tag{32}$$

$$\times \exp\left[\frac{i}{h}[S_0(Q) - S_0(Q')]\right] \rho(Q_\tau Q_\tau')$$

$$\times DQ(t) DQ'(t) dQ_\tau dQ_\tau'$$

with

$$\delta(X_T - X_T') = \sum_f \chi_f^*(T) \chi_f(X_T')$$

The quantity inside the braces is $F(Q, Q')$, for which the final states of the interacting systems have been summed over. Of course, we are not interested in the interacting system. The following result is therefore evident:

$$\rho(Q_T, Q_T') = \int F(Q, Q') \exp\left\{\frac{i}{h}[S_0(Q) - S_0(Q')]\right\} \tag{33}$$

$$\times \rho(Q_\tau, Q_\tau') DQ(t) DQ'(t) dQ_\tau dQ_\tau'$$

8. CLASSICAL POTENTIALS $F(Q, Q')$

If the potential energy term in the Lagrangian is only $V(Q, t)$, no X is involved, and hence

$$F(Q,Q') = \exp\left\{-\frac{i}{h}\int_\tau^T [V(Q,t) - V(Q',t)]dt\right\} \qquad (34)$$

or the influence

$$\Phi(Q,Q') = -\frac{i}{h}\int_\tau^T [V(Q,t) - V(Q',t)]dt \qquad (35)$$

If there are several potentials $\sum_k V_k(Q, t)$, then the influence functional

$$F(Q,Q') = \prod_k F_k(Q,Q') \qquad \sum_k \phi_k(QQ') = \Phi \qquad (35a)$$

Also, if $Q = Q'$ for all t, then for the classical potential $F = 1$. If we take a potential $V(Q, t) = V(Q) C(t)$ where $V(Q)$ is known but the strength $C(t)$ has a probability distribution, we can find the average functional $\langle F \rangle = \sum \omega_r F_r$ where ω_r is the probability for the potential V_r and

$$F_r = \exp\left\{-\frac{i}{h}\int_\tau^T [V_r(Q,t) - V_r(Q',t)]dt\right\} \qquad (36)$$

9. GAUSSIAN NOISE

The potential may be taken as $V(Q, t) = C(t) V(Q)$, where $C(t)$ is random. If $C(t)$ is a Gaussian noise with a power spectrum $\phi(\nu)$ and hence a correlation function $R(\tau) = 2/\pi \int_0^\infty \phi(\nu) \cos \nu t \, d\nu$ we can obtain

$$\langle F \rangle = \left\langle \exp\left\{\frac{i}{h}\int_t^\tau C(t)[V(Q) - V(Q')]\, dt\right\}\right\rangle \qquad (37)$$

Since $C(t)$ is Gaussian,

$$\langle F \rangle = \exp\left\{-(\pi h^2)^{-1}\int_0^\infty \phi_\nu | V_\nu(Q) - V_\nu(Q')|^2 d\nu\right\}$$

where

$$[V_\nu(Q) - V_\nu(Q')] = \int_\tau^T [V(Q) - V(Q')]e^{-i\nu t} dt \qquad (38)$$

10. $F(Q, Q')$ FOR A SIMPLE HARMONIC OSCILLATOR

Consider a test system Q coupled to a simple harmonic oscillator of mass m, characteristic frequency ω, and coordinate X:

$$L_{\text{total}} = L_0(Q, Qt) + \tfrac{1}{2}m\dot{X}^2 - \tfrac{1}{2}m\omega^2 X^2 + QX$$

$$S = S_0(Q) + \int_\tau^T \tfrac{1}{2}m\dot{X}^2 + \tfrac{1}{2}m\omega^2 X^2 + QX)dt \qquad (39)$$

The wave functions of the simple harmonic oscillator are

$$\psi_n(X) = C_n H_n\left[\left(\frac{m\omega}{\hbar}\right)^{1/2} x\right]\exp\left(-\frac{m\omega}{2\hbar}x^2\right)$$

where the H_n are the Hermite polynomials of degree n.

$$\psi_n(X) = 2^{-n/2}(n!)^{-1/2}\left(\frac{m\omega}{\hbar\pi}\right)^{1/4}\exp\left(-\frac{m\omega}{2\hbar}x^2\right)H_n\left[\left(\frac{m\omega}{\hbar}\right)^{1/2}x\right] \qquad (40)$$

If the simple harmonic oscillator is originally in ground state at zero temperature $\chi_i(x) = \exp(-m\omega x^2/2\hbar)$ and since we are not interested in the final state of the oscillators, we have

$$\sum_f \chi_f^*(X_T)\chi_f(X_T') = \delta(X_T - X_T')$$

and hence

$$F_0(Q, Q') = \int \delta(X_T - X_T')K_Q(X_T, X_\tau)K_Q^*(X_T', X_\tau')$$

$$\times \exp\left(-\frac{m\omega}{2\hbar}\right)(X_\tau^2 + X_\tau'^2)dx_\tau dx_\tau' dx_T dx_T'' \qquad (41)$$

$K_Q(X_T, X_\tau)$, as is already known, is the classical K function with the exponential expressed by

$$\exp\left\{\frac{i}{\hbar}[S(X) - S_I(Q, X)]\right\}_{\text{classical}} \qquad (42)$$

that is,

$$K_Q(X_T, X_\tau) = N\exp\left[\frac{i\omega}{2\hbar\sin\omega(T-\tau)}\right]\left[(X_T^2 + X_\tau^2)\cos\omega(T-\tau)\right.$$

$$-2X_T X_\tau + \frac{2X_T}{\omega}\int_\tau^T Q_t\sin\omega(t-\tau)dt$$

$$+\frac{2X_\tau}{\omega}\int_\tau^T Q_t\sin\omega(T-t)dt \left.\right] \qquad (43)$$

$$-\frac{2}{\omega_2}\int_\tau^T\int_\tau^t Q_t Q_s\sin\omega(T-t)\sin\omega(s-t)ds\,dt$$

N is the normalizing constant. The integrals are Gaussian integrals; integrating them gives the influence phase in an equation, from which, equating the imaginary parts, we obtain the required Φ_0:

$$i\Phi_0(Q,Q') = -(2mh\omega)^{-1} \int_\tau^T \int_\tau^t (Q_t - Q'_t) \tag{44}$$
$$\times \{Q_s \exp[i\omega(t-s)] - Q'_s \exp[-i\omega(t-s)]\, dt\, ds$$

or, in general,

$$i\Phi_0(Q,Q') = -\frac{1}{2\hbar} \int_{-\infty}^{+\infty} \int_{-\infty}^t (Q_t - Q'_t)[Q_s F^*(t-s) - Q'_s F(t-s)]\, ds\, dt$$

where

$$F(t-s) = A_0(t-s) + iB(t-s) \quad \text{for} \quad t > s \tag{45}$$

and

$$B(t-s) = \frac{1}{m\omega} \sin \omega(t-s)$$

We shall transform this expression according to the following:

$$Q_\nu = \int Q_t \exp(-i\nu t)$$

Let us take the term

$$\int_\tau^T \int_\tau^t Q_t Q_s F^*(t-s)\, dt\, ds \tag{46}$$

Here, τ and T can be interpreted as turning the coupling on (between X and Q) at $t = \tau$ and off at $t = T$. If, however, the interaction system is taken as part of steady-state environment of Q, it is meaningful to extend these limits to $t = -\infty$ and $T = +\infty$:

$$\int_{-\infty}^\infty \int_{-\infty}^t Q_t Q_s F^*(t-s)\, dt\, ds$$

$$Q_t = \frac{1}{2\pi} \int_{-\infty}^\infty Q_\nu \exp(i\nu t)\, d\nu$$

Hence, $\int_{-\infty}^\infty d\nu \int_{-\infty}^\infty \exp(i\nu t)\, dt \int_{-\infty}^t Q(s) F^*(t-s)\, ds$

$$= \int_{-\infty}^\infty d\nu \int_{-\infty}^\infty dt \int_{-\infty}^\infty H(t-s) F^*(t-s) Q(s)\, ds \tag{47}$$

In Fourier transform theory, we have for the inverse M of $[F(\omega)\, G(\omega)]$:

$$M\{F(\omega)G(\omega)\} = \frac{1}{2\pi} \int_{-\infty}^\infty F(\omega)G(\omega) \exp(i\omega t)\, d\omega$$

$$= \frac{1}{2\pi} \int_{-\infty}^\infty F(\omega) \exp(i\omega t) \left[\int_{-\infty}^\infty g(X) \exp(-i\omega X)\, dX \right] d\omega$$

$$= \frac{1}{2\pi} \int_{-\infty}^\infty g(x) \left\{ \frac{1}{2\pi} \int F(\omega) \exp[i\omega(t-x)]\, d\omega \right\} dx$$

$$= \frac{1}{2\pi} \int_{-\infty}^\infty g(x) f(t-x)\, dx \tag{48}$$

Therefore,

$$\int_{-\infty}^{\infty} \left\{ \int F^*(t-s)\,Q(s)\,H(t-s) \right\} \exp(i\nu t)\,dt$$

$$= Q_{-\nu}\left\{ \int F^*(t)\,H(t)\exp(i\nu t)\,dt \right\}$$

$$= Q_{-\nu}\left\{ \int_{-\infty}^{\infty} A(t)\,H(t) + i\int_{-\infty}^{\infty} B(t)\,H(t)\exp(i\nu t)\,dt \right\} \quad (49)$$

$$\frac{1}{2\pi}\left[\int Q_\nu\,Q_{-\nu}\int_{-\infty}^{\infty} B(t)\,H(t)\exp(i\nu t)\,d\nu \right] \quad (50)$$

The response of a system $r(t)$ is the mate of the admittance of the system $Y(\nu)$:

$$r(t) = \frac{1}{2\pi}\int_{-\infty}^{\infty} Y(\nu)\exp(i\nu t)\,d\nu \quad (51)$$

or $1/Y_\nu = Z_\nu$, the impedence of the system. For the unit step function, the transform is $1/i\nu$. Therefore, the transform of $H(t)\,\mathrm{Im}\,F(t)$ is really $1/i\nu Z$:

$$\frac{1}{i\nu Z_\nu} = \int_0^\infty B(t)\exp(i\nu t)\,dt$$

That is, $\quad \dfrac{1}{m\omega}\displaystyle\int_0^\infty \sin\omega t\,\exp(i\nu t)\,dt$

$$= \frac{1}{m\omega}\int_0^\infty \frac{\exp(i\omega t) - \exp(-i\omega t)}{2i}\exp(i\nu t)\,dt$$

$$= \frac{1}{m\omega}\tfrac{1}{2}\int \exp[i(\omega + \nu + i\epsilon)t]\,dt \quad (52)$$

$$- \frac{1}{m\omega}\tfrac{1}{2}\int_0^\infty \exp[-i(\omega - \nu \quad i\epsilon)t]\,dt$$

The $i\epsilon$ are introduced to make the integrals convergent:

$$-\frac{1}{2m\omega}\left\{ \frac{1}{i(\omega + \nu + i\epsilon)} + \frac{1}{i(\omega - \nu - i\epsilon)} \right\}$$

$$= \frac{1}{im}\frac{1}{(\nu + i\epsilon)^2 - \omega^2} \quad (53)$$

The other term contains $\exp[-i\omega(t-s)]$ and so will lead to the denominator $(1/im)\{1/[(\nu - i\epsilon)^2 - \omega^2]\}$. Since all the quantities are of the form $Q_\nu\,Q_{-\nu}^1$, etc., the integrand is even and $\displaystyle\int_{-\infty}^{\infty}$ is replaced by $2\displaystyle\int_{-\infty}^{\infty} d\nu$. Hence, the $(i\nu Z_\nu)^{-1}$ corresponds to

$$\frac{1}{m\omega}\int_0^\infty \sin\omega t\,\exp(-i\nu t)\,d\nu$$

Now this Z_ν can be shown to correspond to the classical response for the oscillator, an impulse function. Taking the classical oscillator,

$$m\ddot{X} + m\omega^2 X = f \tag{54}$$

This solution corresponds to the initial conditions $X(0) = 0$, $\dot{X}(0) = 0$. Taking Fourier transforms on both sides, we see for $\dot{X}(0) = 0 = X(0)$,

$$-m\nu^2 X\nu + m\omega^2 X_\nu = f_\nu \tag{55}$$

If $f = \delta(t)$, an impulse function, the solution X_ν is

$$X_\nu = \frac{f}{m[\omega^2 - (\nu \pm i\epsilon)^2]} = \frac{1}{-m\{(\nu \pm i\epsilon)^2 - \omega^2\}}$$

Therefore,

$$\Phi(QQ') = \frac{1}{2\pi\hbar} \int_0^\infty \left\{ \frac{Q'_\nu[Q_\nu - Q'_{-\nu}]}{i\nu Z_\nu} + \frac{Q_{-\nu}[Q_\nu - Q'_\nu]}{-i\nu Z_{-\nu}} \right\} d\nu \tag{56}$$

where Z_ν is the classical impedance to an applied impulse force $\delta(t)$ for quiescent initial conditions:

$$f_\nu = \int_0^\infty f(t) \exp(-i\nu t)\, dt \quad X_\nu = \int_0^\infty X(t) \exp(-i\nu t)\, dt$$

that is,

$$X_\nu = \frac{1}{2\pi\hbar} \left\{ \int \left[\frac{Q'_\nu(Q_{-\nu} - Q'_{-\nu})}{-m[(\nu - i\epsilon)^2 - \omega^2]} + \frac{Q_{-\nu}(Q_\nu - Q'_\nu)}{-m[(\nu + i\epsilon)^2 - \omega^2]} \right] d\nu \right.$$

and

$$\frac{1}{i\nu Z_\nu} = \int B(t) \exp(-i\nu t)\, dt \tag{57}$$

or

$$B(t) = \frac{2}{\pi} \int \operatorname{Im}\left(\frac{1}{i\nu Z_\nu}\right) \sin \nu t\, d\nu \tag{58}$$

11. DISTRIBUTION OF OSCILLATORS

Let us think of an interaction system which is a distribution of linear oscillators. Let $G(\Omega)\, d\Omega$ be the weight of the oscillators whose natural frequency is in the range Ω to $\Omega + d\Omega$. Each oscillator is initially in the ground state and finally in an arbitrary state:

$$S[Q, X(\Omega)] = S_0(Q) + \int_\tau^T \int_0^\infty G(\Omega)(\tfrac{1}{2}\dot{X}^2 - \tfrac{1}{2}X^2\Omega^2 + QX)\, d\Omega\, dt \tag{59}$$

We know that if individual disturbances act on Q, the total functional is a product of the individual ones, or the total influence phase is the sum of the individual phases:

$$\Phi_0 = \int_0^\infty G(\Omega)\Phi_{0,\Omega}(Q, Q')\, d\Omega$$

$$\Phi_0 = (2\pi\hbar)^{-1} \int_0^\infty G(\Omega)\, d\Omega \int_0^\infty \left\{ \frac{Q_\nu'(Q_{-\nu} - Q_{-\nu}')}{-[(\nu - i\epsilon)^2 - \Omega^2]} \right. \tag{60}$$

$$\left. + \frac{Q_{-\nu}'(Q_\nu - Q_\nu')}{-[(\nu + i\epsilon)^2 - \Omega^2]} \right\} d\nu$$

Now we can identify the impedance of the system:

$$(i\nu Z_\nu)^{-1} = \lim_{\epsilon \to 0} \int_0^\infty G(\Omega)[(\nu - i\epsilon)^2 - \Omega^2]^{-1}\, d\Omega$$

$$= \frac{1}{Z_\nu} = \frac{\pi}{2} G(\nu) - i\nu \int G(\Omega)(\nu^2 - \Omega^2)^{-1}\, d\Omega \tag{61}$$

$$\left[(\nu - i\epsilon)^2 - \Omega^2]^{-1} = (\nu - \Omega^2)^{-1} + \frac{i\pi}{2\Omega}[\delta(\nu - \Omega) - \delta(\nu + \Omega)] \right]$$

Z has a finite real part, which of course corresponds to a dissipative part, we can arrive at the same impedence classically. Suppose we replace in the Lagrangian, the coupling potential

$$Q(t) \int G(\Omega) X(\Omega, t)\, d\Omega$$

by

$$f(t) \int_0^\infty G(\Omega) X(\Omega, t)\, d\Omega$$

which is a classical potential. The total Lagrangian of the system is

$$L[\dot{X}, X, t] = \int_0^\infty G(\Omega)\, d\Omega [\tfrac{1}{2}\dot{X}(\Omega)^2 - \tfrac{1}{2}\Omega^2 X^2] \tag{62}$$

$$+ f(t) \int_0^\infty G(\Omega) X(\Omega)\, d\Omega$$

The equations of motion of the infinite set are represented by

$$\ddot{X} + \Omega^2 X = f$$

For $X(0) = 0$, $\dot{X}(0) = 0$, and $f(t)$ applied at $t = 0$, we have

$$\frac{X_\nu(\Omega)}{f_\nu} = -[(\nu - i\epsilon)^2 - \Omega^2]^{-1} = [i\nu Z_\nu(\Omega)]^{-1}$$

For the total coordinate X_ν, the ratio X_ν/f_ν is obtained as

$$\frac{X_\nu}{f_\nu} = \int_0^\infty G(\nu)[(\nu - i\epsilon)^2 - \Omega^2]^{-1}\, d\Omega = (i\nu Z_\nu)^{-1} \tag{63}$$

which is the same expression for $(i\nu Z_0)$ obtained in the quantum case. In the classical case, the real part is identified with resistance and the imaginary part with reactance. So, for the case in which the loss is represented by the distribution of oscillators, the effect can be included by using the appropriate impedence function in the influence functional. Suppose there exists a system Q coupled to an assemblage of oscillators, which are also interconnected with each other. Such a linear system of oscillators can be represented by an equivalent set of oscillators (which are the normal modes of the total system) independently coupled to Q. It may be shown that the influence functional for all linear systems has exactly the same form $\exp[i\,\Phi_0(qq')]$, where Φ_0 is a quadratic functional of Q and Q' adapted to a linear system only through the classical response of that linear system to a force. When the coupling Lagrangian is linear in the coordinate X of the oscillator and the test system Q, the elimination of X yields a phase quadratic in Q. This is because eliminating X of the oscillator is nothing but doing a path integral which is a Gaussian integral, and completing the square yields a quadratic term.

12. LINEAR SYSTEMS AT FINITE TEMPERATURES

Let us take a system Q, acted on by a simple oscillator. We set up the problem in exactly the same fashion as in the zero-temperature case; however, the initial state of the linear system is not a ground state or any definite eigenstate, but is uncertain. So it has to be summed over all states weighted by the Boltgmann factor $\exp(-\beta E_n)$. The final state is again arbitrary, as in the previous case. Therefore, we write

$$F(Q,Q') = \int \sigma(X_T - X_{T'})K_Q(X_T, X_{T'})K_{Q'}(X_{T'}, X'_\tau)$$

$$\times \sum \frac{\phi_n(X_\tau)\phi_n^*(X'_\tau)\exp(-\beta E_n)}{\sum_n \exp(-\beta E_n)} \, dX_\tau \ldots dX'_\tau$$

$$(64)$$

We will have to find the sum

$$\sum_n \phi_n(X_\tau)\phi_n^*(X'_\tau)e^{-\beta E_n}$$

Let us take the kernel which takes the oscillator from X_τ to X_τ' in time interval $t_2 - t_1$. This will yield a propagator function of the form

$$K_0(X_2, X_1) = \sum \phi_n(X_2)\phi_n(X_1) \exp\left[-\frac{i}{\hbar} E_n(t_2 - t_1)\right]$$

the ϕ_n being the energy eigenfunctions. This we know is nothing but the kernel $= \exp[i/\hbar\, S_{cl}]$. For the harmonic oscillator

$$S_{cl} = im\omega[\hbar 2 \sin \omega(t_2 - t_1)]^{-1} \qquad (65)$$
$$\times [(X_1^2 + X_2^2)\cos \omega(t_2 - t_1) - 2X_1 X_2]$$

If we make the correspondence $\beta = i(t_2 - t_1)/\hbar$, we see that

$$\sum \phi_n(X_\tau)\phi_n^*(X_\tau') \exp(-\beta E_n)$$
$$= \exp\{-m\omega[2\hbar \sin h(\beta\hbar\omega)]^{-1} \qquad (66)$$
$$\times [(X_\tau^2 + X_\tau'^2)\cos h(\beta\hbar\omega) - 2X_\tau X_\tau']$$

Using this, we find the influence functional to be

$$i\phi(Q, Q') = i\phi_0(Q, Q') - (\pi\hbar^2)^{-1}\int_0^\infty \pi\hbar[2m\omega(e^{-\beta\hbar\omega} - 1)] \qquad (67)$$
$$\times \delta(v - \omega)|Q_v - Q_v'|^2\, dv$$

The first term is the effect at zero temperature. For a Gaussian potential, we find

$$\langle F \rangle = \exp\left[-(\pi\hbar^2)^{-1}\int_0^\infty \phi(v)|V_v(Q) - V_v(Q')|^2\, dv\right]$$

where ϕ_v is the power spectrum of the Gaussian potential. Hence, we see that the effect of temperature is to introduce another term in the phase function which corresponds to a Gaussian noisy potential acting on Q, at the frequency of the original oscillator; the power spectrum of the noise produced by the temperature is given by

$$\phi_v = \pi\hbar[2mv(e^{-\beta\hbar\omega} - 1)]^{-1}\delta(v - \omega) \qquad (68)$$

Suppose we extend this to the case of distribution of oscillators. We obtain for the influence phase

$$\phi(Q, Q') = i\int_0^\infty \phi_0(Q, Q')G(\Omega)\, d\Omega \qquad (69)$$
$$+ (\pi\hbar^2)^{-1}\int \pi\hbar G(v)[2v(e^{\beta\hbar v} - 1)]^{-1}|Q_v - Q_v'|^2\, dv$$

The first term is the influence phase at zero temperature. The second term has the form of a noisy potential whose power spectrum is

$$\phi(v) = \left(\frac{\hbar\pi}{2}\right)G(v)[v(e^{\beta\hbar v} - 1)]^{-1} \qquad (70)$$

From the analysis of distribution of oscillators, it is found that

$$\frac{\pi G(\nu)}{2} = \text{Re}\left(\frac{1}{Z_\nu}\right) \tag{71}$$

Therefore, the power spectrum can be written as

$$\phi(\nu) = \hbar\,\text{Re}\,(Z_\nu)[\nu(e^{\beta\hbar\nu} - 1)]^{-1} \tag{72}$$

We conclude this by applying this result obtained above to obtain Nyquist's result for noise from a resistor. Take an arbitrary circuit as a test system connected to a resistor $R_T(\nu)$ at temperature T which is our interaction system. Let us associate $V(t)$, the voltage across $R_T(\nu)$, as the coordinate of the interaction system and $Q(t)$, the charge flowing through the system, as the coordinate for the test system. The interaction part of the Lagrangian is $-Q(t)\,V(t)$. The quantity $i\nu\,Z\nu$ appearing in the influence functional is

$$\frac{-Q_\nu(t)}{V_\nu(t)} = -(i\nu R_\nu)^{-1} = i\nu Z_\nu$$

Therefore, the real part $\text{Re}(Z\nu)^{-1} = \nu^2 R_\nu$. Accordingly, the random classical voltage across the resistor has a power spectrum

$$\phi(\nu) = \hbar\nu R_\nu(e^{\beta\hbar\nu} - 1)^{-1} \tag{73}$$

The mean value of this voltage is

$$\langle v(t)^2\rangle \frac{2}{\pi}\int_0^\infty \phi(\nu)\,d\nu \tag{74}$$

For high temperature, $\beta \ll 1$ and so $\beta\hbar\nu \ll 1$. For a given frequency range $\phi(\nu) = kTR_\nu$. If R_ν is constant over a given frequency range $(F_2 - F_1)$, then

$$\langle v^2(t)\rangle = 4kTR\,(f_2 - f_1) \tag{75}$$

This is the famous Nyquist result for the noise from a resistor.

REFERENCES

1. R.P. Feynman and F.L. Vernon, Jr., *Ann. Phys.* **24**, 118 (1963).
2. R.P. Feynman, *Rev. Mod. Phys.* **20**, 367 (1948).

Author Index

Subject Index